Savanna Burning

Understanding and using fire in northern Australia

Edited by

Rodd Dyer
Peter Jacklyn
Ian Partridge
Jeremy Russell–Smith
Dick Williams

Contents

6. Burning operations   81

7. Monitoring fire regimes   102

8. Global trends and fire management   122

Glossary   130

Index   132

Acknowledgements

The editors would like to thank:

All chapter contributors and funding bodies including Meat and Livestock Australia, Natural Heritage Trust and Bushfires Council of the NT for their support

State of the Environment Australia and WA Department of Land Administration (DOLA) for use of fire-mapping products

Terry Underwood, Russell Anderson and Phil Cheney for their reviews and for providing useful comments in the final stages

Ian Partridge*, Queensland Department of Primary Industries, for the initial editing, much of the layout design and for his professional expertise in polishing the text

Deb Bisa   for her thorough proof-reading

John Merrick for compiling the comprehensive Index

Dick Fell for putting together the information in case studies

Skyscans, Mick Todd, BFCNT, DOLA(WA), DPIF and Terry Mahney for the cover photos

All other suppliers of photographs

Simon Love of Stingray Advertising and Design, Darwin for his patience, assistance with layout development and excellent graphic design

And finally the Cooperative Research Centre for Tropical Savanna Management and its CEO John Childs for providing a forum where this enjoyable and productive collaborative process could be undertaken.

Chapter authors

Grant Allan ~ Bushfires Council of the NT

Garry Cook ~ CSIRO Sustainable Ecosystems

Peter Cooke ~ Caring for Country Unit, Northern Land Council

Rodd Dyer* ~ NT Department of Primary Industry and Fisheries

Tony Grice ~ CSIRO Sustainable Ecosystems

Peter Jacklyn* ~ Tropical Savannas CRC

Andrea Johnson ~ NT Department of Primary Industry and Fisheries and Tropical Savannas CRC

Tim McGuffog ~ Bushfires Council of the NT

Jeremy Russell-Smith* ~ Bushfires Council of the NT and Tropical Savannas CRC

Tony Start ~ WA Department of Conservation and Land Management

Dick Williams* ~ CSIRO Sustainable Ecosystems

Brent Williams ~ Bushfires Council of the NT

John Woinarski ~ Parks and Wildlife Commission of the NT

Cameron Yates ~ Bushfires Council of the NT

Dean Yibarbuk ~ Bawinanga Association

Other contributors

Alan Andersen ~ CSIRO Sustainable Ecosystems

Andrew Ash ~ CSIRO Sustainable Ecosystems

Shane Campbell ~ Queensland Department of Natural Resources and Mines

Andrew Craig ~ Agriculture Western Australia

Gabriel Crowley ~ Queensland National Parks and Wildlife Service

Andrew Edwards ~ Bushfires Council of the NT

Rod Fensham ~ Queensland Herbarium

Don Franklin ~ Key Centre for Tropical Wildlife Management, NTU

Max Gorringe ~ Manager, Elsey Station

Tony Griffiths ~ Key Centre for Tropical Wildlife Management, NTU

Trevor Howard ~ Bushfires Council of the NT

Roger Landsberg ~ Manager, Trafalgar Station

Darrell Lewis ~ History Consultant

Jeff Little ~ Manager, Opium Creek Station

Lyn Lowe ~ CSIRO Sustainable Ecosystems

David Phelps ~ Queensland Department of Primary Industries

Gary Riggs ~ Former manager, Mataranka Station

Lisa Roeger ~ Bushfires Council of the NT

Tom Starr ~ Manager, Ban Ban Springs Station

Peter Stanton ~ Ecological Consultant

Tom Stockwell ~ Manager, Sunday Creek Station

Simon Townsend ~ NT Department of Lands, Planning and Environment

Peter Whitehead ~ Key Centre for Tropical Wildlife Management, NTU

***Editors**

Overview

This book describes how fire affects all aspects of the savanna landscape; it also provides some guidelines on how fire can be used to achieve multiple aims. It is designed as a concise, readable account of our current knowledge, and also as a general reference relating to fire management in the savannas of northern Australia.

Book overview

To help you see which parts of the book might be most useful the chapters are outlined below.

Chapter 1. Introduction

The Introduction describes the fire management issues that need to be addressed in northern Australia and introduces the people will need to be involved in solutions to these issues.

Chapter 2. Savanna landscapes

Savanna Landscapes describes the physical environment of the tropical savannas and shows how the different landscapes within the savannas have been shaped by climate, geology and soils. Despite this variation, all savanna landscapes share similar fire issues. Indeed these landscapes have been shaped by fire and long-term patterns of land use. Present day land use and fire patterns are in turn shaped by the landscapes and climate of the north.

Chapter 3. Savanna fire regimes

Having been given the physical setting, this chapter goes on to describe the different types of fire in the tropical savannas in more detail and introduces the concept of fire 'regimes'. It describes the broad characteristics of tropical savanna fire regimes and shows how they are affected by weather and vegetation—or 'fuel'. This chapter shows how different fire regimes are used by different land users for various tasks in the landscape. It briefly describes the current problems with the present fire regimes and how they affect conservation, production and culture.

Chapter 4. Effects of fire in the landscape

This chapter details the effects of the current fire regimes on the landscape—how the current fire patterns are impacting on activities and values ranging from pastoralism to traditional practices to biodiversity. How are fire regimes affecting the ratio of trees to shrubs; of annual to perennial grasses? How are they affecting fire-sensitive plant species in sandstone country? What impact do fire patterns have on various native animals? How do they change air quality and water quality?

Chapter 2 describes the tropical savannas in physical terms.

Chapter 4 details how fire patterns impact on biodiversity.

Chapter 5. Using fire in savanna management

Using fire shows how better fire management can overcome many of the problems described in Chapter 4. Fire management techniques are described for Aboriginal land, pastoral land and for managing biodiversity. It is explained, for example, which burning regimes can be used to enhance pasture vigour; to control weeds; to provide habitat for native animals and to protect against wildfires.

Chapter 6. Burning operations

Once you know what fire regimes are needed, 'operations' describes how to put that knowledge into practice. It describes when and where to light fires for particular tasks; what ignition techniques can be used and how to plan for burning. The construction of firebreaks is detailed. The legislation that governs the lighting of fire in WA, NT and Queensland is also referenced.

Chapter 7. Monitoring fire regimes

How do you know that fire management is working? You need to be able to monitor the effects of fire on the landscape and then, if necessary, use these results to change existing fire management. Monitoring techniques involving plots, aerial photography and satellite remote sensing are described. The complex language and high technology of satellite monitoring is explained.

Chapter 8. Global trends and fire management

Finally, we look to the future. This chapter examines how fire management will be affected by various global and local trends including greenhouse issues and carbon trading, globalisation of markets, new laws covering biodiversity and air quality, and the impact of native title legislation.

Chapter 5 shows how to protect against wildfires.

Chapter 8 discusses global trends and issues.

Preface

by John Childs, CEO Tropical Savannas CRC

For thousands of years, a combination of change to drier climates and Aboriginal land management has resulted in fire shaping Australian landscapes. There have been major changes over the past 100–150 years with European attitudes and approaches replacing those of Aboriginal people on many lands. The consequence has been changes in landscape structure and habitat, and in the frequency of catastrophic and destructive fires.

Over the past few years our knowledge and understanding of fire has advanced considerably. We know where fire occurs and how frequently. We better understand the impact of fire on different plant communities, fauna populations and cattle grazing behaviour. And we have a range of methods and tools to help us determine when and where fires occur.

We are now able to develop strategies for the use of fire to shape the landscape in ways that maintain their health and productivity. This can be applied whatever the use being made of the landscape. We can also develop strategies which avoid or reduce the impact of fire in its more destructive forms.

This book is a collection of the current state of knowledge concerning fire and burning in northern Australia. It incorporates the knowledge and work of many people engaged both within Tropical Savannas CRC and in other agencies and enterprises. It also provides support and assistance to those wishing to use fire in managing the landscape.

The information and knowledge in this book provides the basis for future research, application, education and communication related to fire and its place in the northern savanna landscapes. We still have much to learn. We will achieve this through the thoughtful application of the knowledge presented in this book and reviewing the consequences and impacts.

I would particularly like to pay tribute to Jeremy Russell-Smith for his inspiration and guidance in promoting the importance of fire in the landscape and developing many of the significant research and development activities being undertaken. He has been a driving force in Tropical Savannas CRC in the development of our program and in the production of this book.

This book and the knowledge it represents are a step along the way to a better understanding and use of fire in northern savanna landscapes. Further progress will be made by people using and studying fire in the landscape and communicating what they learn. We encourage you to join us in this endeavour.

John Childs, CEO Tropical Savannas CRC

1. Introduction

by Rodd Dyer, Peter Jacklyn, Jeremy Russell-Smith and Dick Williams

North Australia has the largest and most frequent fires in the whole continent (Figure 1.1). This part of the country is primed for regular fire with vast grassy landscapes that flourish during the wet season and then, over months with very little rain, dry to a tinderbox.

The problems these fires create for the people of the north are considerable, whether it is in managing cattle stations or protecting fire-sensitive plants and animals. Despite extensive research, practical information on how to manage fire in the north is often hard to come by, yet sorely needed. This book aims to fill that need.

The issues

In the higher rainfall savanna woodlands—in the north Kimberley, the Top End and Cape York—as much as half of the area is burnt either every year or every second year, typically late in the dry season (as shown in the red areas in Figure 1.1). As these late fires are generally intensely hot and extensive in area, they have the potential to devastate populations of fire-sensitive native plants and animals, to be costly and disruptive to pastoral operations, to pose a threat to communities and property, while having implications for greenhouse gas emissions.

Further south, east and inland, fires become less frequent because of the lower annual rainfall and generally more intensive use of savannas for grazing cattle. This reduces the fuel available for fires. Wildfires are often actively suppressed, and prescribed burning is generally excluded so that pasture can be used as livestock forage rather than as fuel for fire.

In many of these inland regions, there is evidence that this reduced burning has contributed to the unchecked growth and increasing dominance of native trees and shrubs in once open grasslands and woodlands.

What has changed?

These newer fire patterns differ from those of traditional Aboriginal management. With traditional practices, small patches of grassland were burned throughout the year as people moved through the country. In the Top End and northern Kimberley, many fires were lit in the early to mid-dry season. Smaller and less intense fires of traditional practices maintain more diverse habitats than large late fires.

Figure 1.1 Fire map, northern Australia 2000

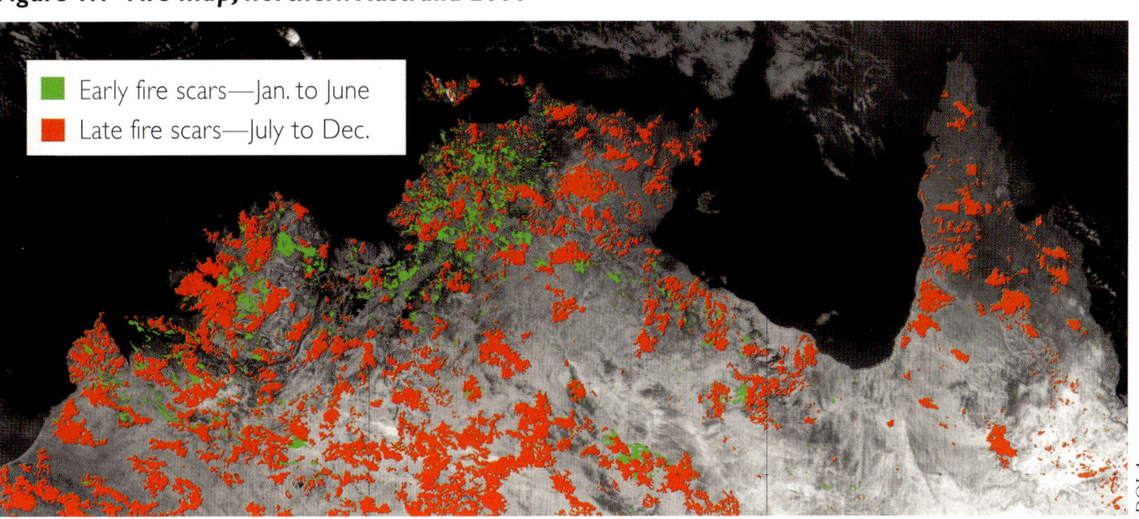

Different land users, different needs

Land use in the northern Australian savannas has also changed—it is now a patchwork of pastoral, conservation, indigenous, military and mining activities (Figure 1.2). While mining and, increasingly, nature-based tourism, are the major economic drivers, pastoralism is by far the most extensive land use. All these savanna land managers will play different roles in managing fire effectively, but need to appreciate other land users.

Pastoralism

Grazing lands are usually made up of large cattle stations, but with some sheep properties in parts of central Queensland. Properties range in size from 300 to 12,000 sq. km; they tend to be smaller on more fertile soils and larger in the more arid regions (Figure 1.3). Individual paddocks can range in size between 10 and 2000 sq. km, and are usually a patchwork of different vegetation, soil and landforms. These pastoral properties need to be run efficiently on relatively little capital and labour. Fire can be an effective tool on pastoral leases with strategic burning being used to prevent wildfire, improve pasture, manage grazing, control weeds and enhance biodiversity.

Figure 1.2 Land use in northern Australia

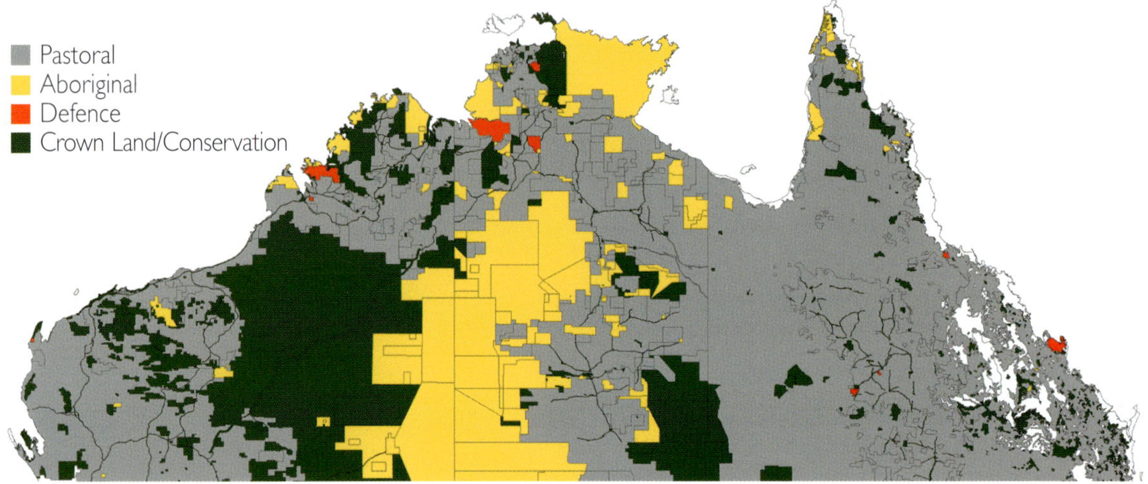

Figure 1.3 Cadastral map

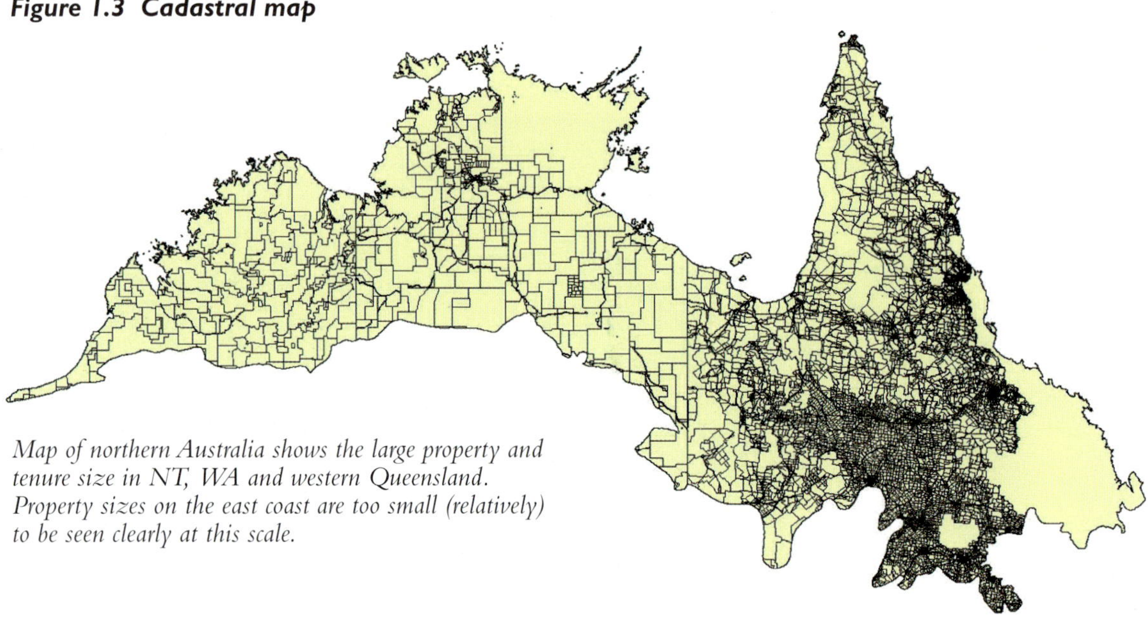

Map of northern Australia shows the large property and tenure size in NT, WA and western Queensland. Property sizes on the east coast are too small (relatively) to be seen clearly at this scale.

Aboriginal lands

Aboriginal communities in north Australia play an important role in fire management as the lands they manage are extensive and indeed are expanding through native title claims and purchase of pastoral leases. These communities have a profound knowledge of fire management that has economic, cultural and spiritual significance. This knowledge base, however, is becoming increasingly precarious as traditional elders age and die.

Conservation

The landscapes of northern Australia are now an important refuge for the country's biodiversity that has international significance. Although little of the savanna lands has been cleared, there are early warning signs of a decline in the abundance and richness of this biodiversity—and evidence that inappropriate fire regimes may be partly responsible. Understanding how fire affects biodiversity therefore has national significance. The importance of conservation outside reserves is increasingly being recognised, and fire has a role to play in managing biodiversity on these lands.

The diverse users of the tropical savannas will play different roles in managing fire effectively, but need to appreciate the other aims and points of view.

Tourism

A number of natural landscape 'icons' in the north, such as Kakadu, the Kimberley and Cape York, make northern Australia a popular destination for tourists. However, tourists, particularly from more temperate climes, may be surprised and confused at the first sight of burnt country. This can result in conflict over perceptions about what is 'best' for the country.

Defence

The military also has a major presence in northern Australia for training both army and airforce. The defence forces have a major stake in fire management as their land may be fire prone, being unstocked and subject to incendiary devices. Military lands also have to be managed in accordance with a number of Commonwealth statutes, such as endangered species legislation.

Mining

Mining operations impact on a small area of the savannas. However, mining exploration and mine site rehabilitation both involve fire management.

The need for fire management information

While individual land managers will be interested in different aspects of fire management, none operate in isolation. All should be aware of all the effects of fire when applying sound fire management practices.

Over the last decade, state, territory and Commonwealth agencies, the CSIRO, and funding bodies such as Meat and Livestock Australia (MLA) and the Commonwealth's Natural Heritage Trust (NHT), have provided essential funds and resources for investigating a range of fire management issues on pastoral, conservation and Aboriginal lands. Until now, much of this new knowledge and technology has existed only in research papers and reports, making it difficult for land managers to access and gain benefit.

2. Savanna landscapes

by Dick Williams and Garry Cook

What are savannas?

Savannas are grassy landscapes—woodlands with a grassy ground layer, or grasslands—that occur in tropical areas where the climate is seasonally dry.

Savannas are important nationally and internationally. They cover about 20% of the world's land surface and about 20% of the Australian continent. The many different types of savanna we see in Australia reflect broad differences in rainfall and soil patterns, but the basic climate of the savannas makes them all prone to fire and frequent fire in particular. Hence, to understand and manage savannas we need to understand not only how climate and soil influence them, but how fire has shaped, and will continue to shape, the savanna landscapes.

In Australian savannas, the trees are mostly eucalypts, but several species of tall shrub may occur in association with the eucalypts. The grasses and other herbs may be annual or perennial, and the grass layer is usually dense. Tree and shrub 'canopy cover' (the proportion of ground covered or shaded by the tree's canopy) may vary between <1% and 60–70%, and tree density may be up to 100 per hectare. But even in the denser woodlands, the canopies of eucalypts do not generally overlap, and light intensities beneath the canopies are relatively high. Thus, competition between trees, shrubs and grasses is generally for water and nutrients rather than sunlight.

Fire has played an important role in the evolution of savanna landscapes.

The landscapes

Evolution

Australian savannas have been evolving for many millions of years. During the early Tertiary period (about 60 million years ago), the continent was much warmer and wetter than at present and rainforests were much more extensive. Since then, the continent has become drier, the rainforests have contracted, and savannas have expanded. The extensive savannas and warm, wet–dry climate, key features of northern Australia, have existed for at least the last 10 million years. The broad geological features of northern Australia are also many millions of years old.

Because of a long history of evolution, savannas are highly complex and diverse ecosystems. Although they have a simple structure, savannas are rich in species, communities of plants and animals, and habitats. In addition to high species richness, there are numerous life forms (e.g. trees, shrubs, grasses, sedges, herbs, vines), and the savanna plants vary enormously in the ways they cope with life in a strongly seasonal climate.

The savannas also have many different animal species, both vertebrate and invertebrate. Indeed, savannas are richer in many plant and animal groups than are the monsoon rainforests.

Fire has influenced the nature of the savannas over the course of their evolution. Fire has become more frequent as the continent has dried out. Indigenous people have used fire in the savannas for tens of thousands of years, and people continue to use fire for many types of land management purposes. However, there have been many changes to the fire regimes—the extent, frequency, severity and timing of fires—over evolutionary, prehistoric and contemporary times.

Landforms

Most of the savanna region is less than 500 m above sea-level, with local relief generally less than 100 m. Soil types may vary considerably, and the current structure and composition of savannas reflects variations in annual rainfall and in soil texture.

In both higher rainfall and drier savannas there are three broad landforms:

- flat to hilly savanna woodlands
- 'stone country'
- 'black soil plains'.

Within each, there are also 'riparian areas' along river and stream banks; there are important components of the savannas.

Most of the savanna woodlands are flat to gently undulating—the 'lowland plains' country—but there are steeper ranges associated with the Great Dividing Range in eastern Queensland.

The 'stone country' consists of rocky escarpments, slopes and plateaux, and occurs, for example, in eastern Arnhem Land and parts of the Victoria River District in the NT, the Kimberley in WA and north-eastern Cape York in Queensland.

The 'black soil plains', with their cracking clays, are found on the geologically recent flood plains of the major river systems in the wetter regions of the savannas, and on more recent (Quarternary) deposits, or older (pre-Tertiary) fine sediments and basalts in the semi-arid savannas. The riparian areas usually have a narrow, but relatively dense band of trees, often with a grassy understorey.

The soils include sands, loams (e.g. red and yellow earths) and heavy cracking clays (black soils). Each may be locally extensive; loams and clays tend to be more common in the drier savannas, although sands and loams are more extensive in the wetter savannas. The soils of the stone country tend to be shallow and poorly developed, although deep sandy soils may occur on the outwash slopes at the base of slopes and plateaux.

Flat to undulating savanna woodlands

Stone country

Black soil grassland plains

Riparian vegetation

The climate of the savannas

The key feature of the savanna climate is the alternating hot wet and warm dry seasons. This is driven by the monsoon as it moves between the northern and southern hemispheres. The dramatic seasonal variation in climate has a great impact on fire weather through the year (see Chapter 3), and hence on fire regimes.

Savannas may be relatively wet (mesic) where annual rainfall is greater than about 900 mm, or semi-arid/arid where it is 300–900 mm. The long-term average annual rainfall is strongly related to latitude over most of the savanna region (Figure 2.1). In the Top End, Kimberley and Cape York Peninsula, average annual rainfall declines sharply with increasing latitude, in places at a rate of 1 or 2 mm/km south. Rainfall also declines very sharply from east to west on the eastern seaboard near Cairns and Townsville.

The onset and duration of the wet season varies considerably between years and regions, and the amount of winter rain varies between the east and west of the savanna region. Rain generally commences during September–October. It has usually ceased by May in the western Top End of the NT and Kimberley regions, but may continue well into June further to the east of the continent (Figure 2.2). In general, the dry season starts rapidly, with sharp declines in both humidity and surface soil moisture.

Most rainfall in the savannas comes from the monsoonal troughs and/or from isolated convective storms. In both the wet and semi-arid savannas, cyclones may also be a source of heavy rainfall. The influence of the monsoon is stronger in the near-coastal areas than in inland areas (Figure 2.3).

Figure 2.1 Rainfall map of northern Australia

Figure 2.2 Seasonal variations

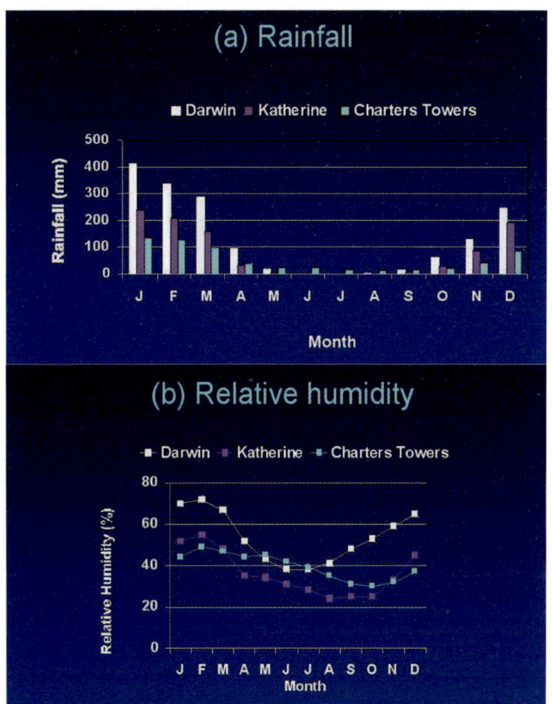

(a) average monthly rainfall and (b) 3 p.m. relative humidity at Darwin (NT), Katherine (NT) and Charters Towers (Qld). Note that the savannas of the eastern seaboard receive more winter rain than do those of the north-west of the continent.

Figure 2.3 Length of the wet season

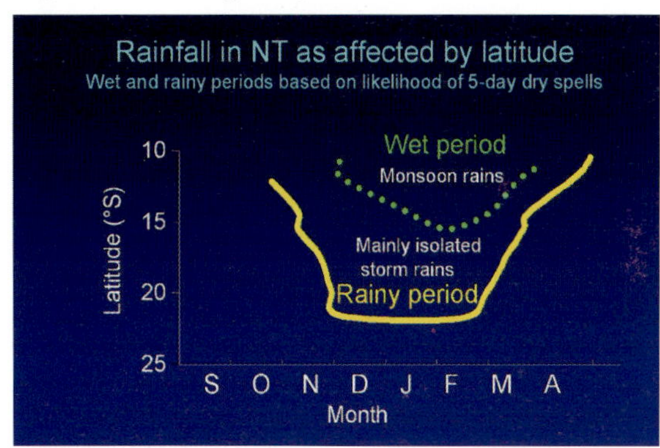

The length of the wet season in the NT increases with decreasing latitude. Rain comes from both the monsoon and storms in the north, but generally only from storms further south and inland.

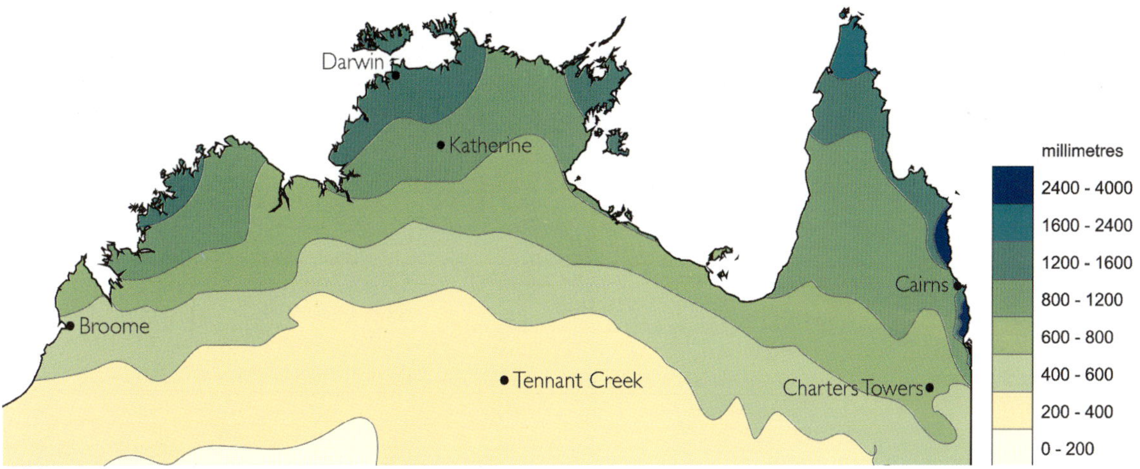

Total rainfall declines with increasing distance from the coast and increasing latitude.

In the Top End of the NT for example, the strong monsoonal influence during the wet season around Darwin and Katherine (the 'wet period' of Figure 2.3) means that there is not only more rain than the more inland areas, such as in the Elliot/Tennant Creek region, but that the wet season (the 'rainy period' of Figure 2.3) is also longer.

Aboriginal people further subdivided both the wet and the dry seasons. In Kakadu, for example, they recognised the 'build-up' period between dry and wet seasons, the wet was divided into the early storm and full monsoon period and there was the period of 'knock-em down' storms at the very end of the wet season. The dry was divided into the early-cool and late-hot dry seasons.

Figure 2.4 Pasture types of northern Australia (Tothill and Gillies 1992)

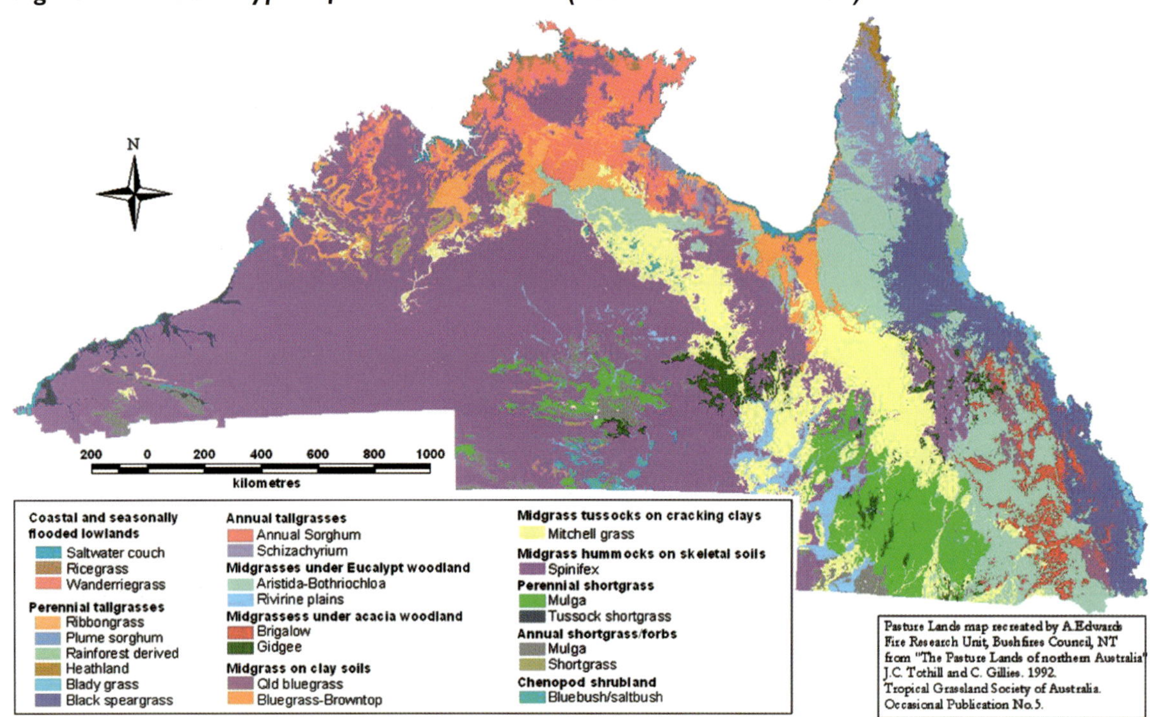

Figure 2.5 Generalised northern Australia vegetation map

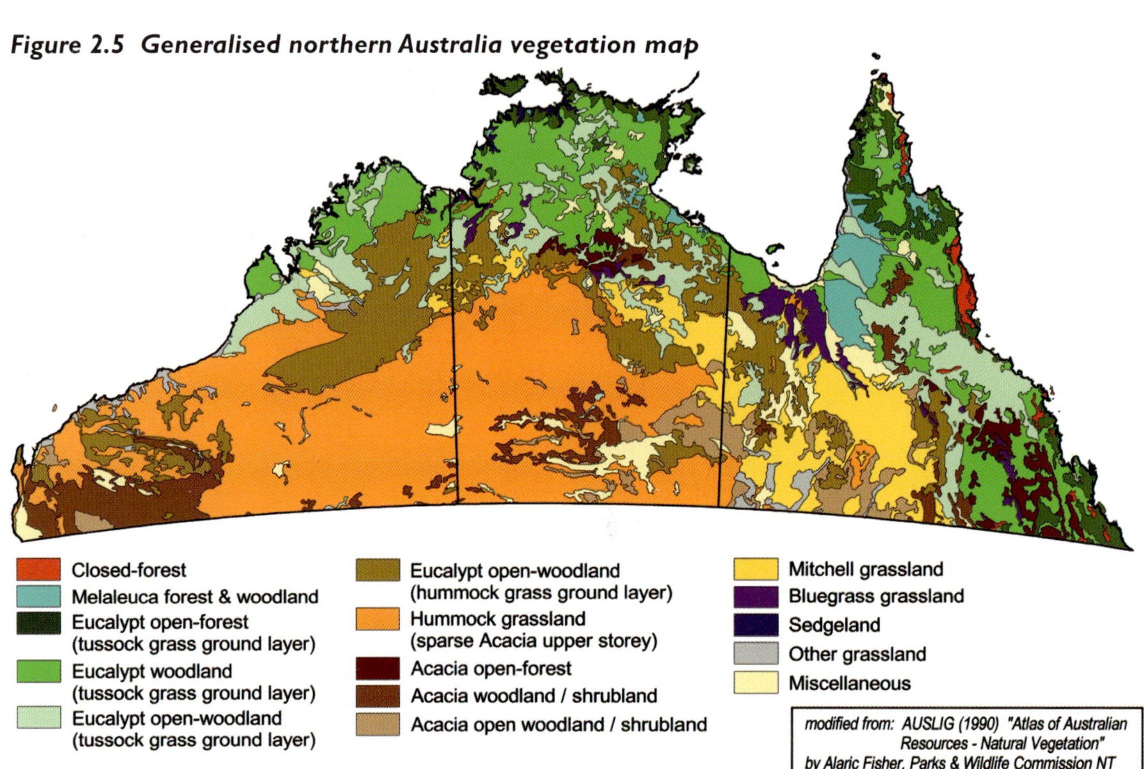

The savanna vegetation types

The many and varied savanna types reflect both the amount of annual rainfall and the soil types. Variation in these two key factors determines both the dominant trees of a savanna and the associated combination of grasses (the 'pasture types') in the understorey.

The savanna types have been mapped extensively, for example the 'Land System Maps' of the 1950s and 1960s, and the maps of the different 'pasture types' (Figure 2.4). Recently, a generalised vegetation map of northern Australia has been produced (Figure 2.5).

Monsoon patch of vine forest

Sandstone habitat

Riparian corridors adjoining watercourses

Variation and diversity in savannas

The savannas range from 'open forest' in the coastal and sub-coastal regions to 'woodlands' in the semi-arid regions to 'open woodlands' with scattered low trees in the arid interior (Figure 2.5). Treeless grasslands occur on heavier soils and where drainage is impeded. Most ecosystems in northern Australia are 'grassy landscapes'. Notable exceptions are the rainforests (the Wet Tropics in north Queensland and the monsoon forests and vine thickets of Queensland, Top End and Kimberley), some of the wetland ecosystems and the most rugged and rocky landscapes of the Kimberley and western Arnhem Land where the vegetation is sparse scrub or heath. These environments make an important contribution to both plant and animal biodiversity of northern Australia, and the health and management of these landscapes cannot be readily disentangled from that of the surrounding savanna. Variation in the structure of the savannas of the north is illustrated in the series of photographs on pages 10 and 11.

Forests and woodlands

Most Australian savannas are eucalypt open forests and woodlands, but they vary considerably because of variation in annual rainfall and soil texture. Broad-leaf trees and shrubs such as *Terminalia* (billy goat plum), *Brachychiton* (kurrajong) and *Erythrophleum* (ironwood) may also occur together with the eucalypts in both the wetter and drier savannas.

In the higher rainfall areas on the lighter textured soils, open forest is typically dominated by eucalypts 10–20 m tall with a canopy cover of 40–60%. The understorey consists of annual or perennial tall grasses. In the Kimberley, Top End and Cape York, such forests are dominated by *Eucalyptus tetrodonta* (stringybark) and *E. miniata* (woollybutt); open *Melaleuca* forests and some patches of monsoonal vine forest fringe the treeless flood plains. Heaths, with a few trees, and hummock grasses (spinifex) grow on the 'stone country'.

In the semi-arid regions, the savannas are woodlands and open woodlands. The eucalypts are shorter (5–15 m tall) and have lower cover (5–30%) than those of the forests. There are numerous species that dominate, but common ones in north-western Australia are bloodwoods and boxes, e.g. *E. tectifica* (grey box), *E. pruinosa* (silver box) *and Corymbia terminalis* (bloodwood). Common species in north Queensland include the iron-barks and boxes, e.g. *E. crebra* (narrow-leaved ironbark), *E. melanophloia*

Open forest

Open forest of woollybutt with annual Sorghum *understorey, Kakadu, NT*

Open forest

Open forest of woollybutt with annual Sorghum *understorey, Katherine, NT*

Woodland

Woodland of bloodwood over annual Sorghum *and perennial grass, Kakadu, NT*

Woodland

Wiregrass under lancewood (Acacia shirleyii), *Daly Waters, NT*

Woodland

Eucalypt woodland (bloodwood) over tall perennial grass ('Tippera' system), Katherine, NT

Woodland

Wiregrass/firegrass under eucalypts, Boroloola, NT

Woodland

Black speargrass under eucalypts, Charters Towers, Qld

Woodland

Wanderrie grass under eucalypts, Normanton, Qld

Open woodland

Eucalypt woodland with arid, short-grass understorey, VRD, NT

Open woodland

Eucalypt and bauhinia (Lysiphyllum) *woodland with annual* Sorghum, *Katherine, NT*

Open woodland

White grass under eucalypts and Terminalia, *Kalkarindji, NT*

Grassland

Ribbon grass grassland adjacent to eucalypt and Terminalia *woodland, VRD, NT*

Open woodland

Ribbon grass under boab trees, Ord River, WA

Open woodland

Spinifex (Triodia) *under snappy gums, Cloncurry, Qld*

Grassland

Barley Mitchell grass (Astrebla) *on inland black soil plains, Kalkarindji, NT*

Grassland

Bluegrass-browntop on coastal black soil plains, Burketown, Qld

(silver-leaved ironbark) and *E. brownii* (Reid River box), and bloodwoods, e.g. *Corymbia erythrophloia* (red bloodwood). Perennial grasses here include kangaroo grass (*Themeda triandra*), black spear grass (*Heteropogon contortus*), ribbongrass(*Chrysopogon fallax*) and white grass (*Sehima nervosum*) on the lighter textured soils. On the poorest and most shallow soils in the lower rainfall areas, tree cover is sparse (1–2%) and spinifex (*Triodia* spp.) predominates.

Some *Acacia*-dominated woodlands provide the exception to the rule of eucalypt predominance in northern Australia. Notable examples were the extensive areas of brigalow open forest (*Acacia harpophylla*) on clay soils in eastern Queensland, large areas of which have been cleared and sown with pasture grasses such as buffel (*Cenchrus ciliaris*) and rhodes (*Chloris gayana*); lancewood (*A. shirleyi*) on lateritic soils of the Sturt Plateau in the Northern Territory, and pindan (*A. eriopoda* and *A. tumida*) on sandy soils in the Dampierland region of Western Australia. Woodlands and open woodlands dominated by gidgee (*A. cambagei* and *A. georginae*) also occupy substantial areas of fine-textured soils in central and western Queensland.

The grassy understoreys also vary with rainfall and soil. Annual sorghums (also often called 'speargrass') are common beneath the eucalypt woodlands and open forests of the wetter areas of the western Top End of the Northern Territory and parts of the Kimberley. In eastern Arnhem Land and Cape York, in contrast, annual *Sorghum intrans* is uncommon, and perennial Sorghum (*S. plumosum*), firegrass (*Schizachyrium fragile*) and other tall grasses predominate.

In the semi-arid savannas, perennial grasses predominate, with common species being ribbongrass, white grass, kangaroo grass and black speargrass. Black speargrass communities occur on free-draining duplex soils (sand over clay), and cover a huge area of eastern Queensland. As rainfall declines westward in north Queensland, *Aristida*, *Bothriochloa* and *Chrysopogon* spp. begin to dominate. In the driest savannas of northern Australia, where soils are skeletal, the grass layer is dominated by 'spinifex', under either open eucalypt woodland or mulga (*Acacia aneura*). Unlike other perennial grasses, which have a tussock growth form, spinifex has a hummock growth form.

Grasslands

Under some conditions, trees are absent and the savannas are grasslands. The most extensive of these grasslands occur in the semi-arid regions on cracking clay soils, such as on the Barkly Tableland, and are dominated by Mitchell grasses (*Astrebla* spp.).

Savanna woodlands become less dense as average rainfall declines. **Top**—*coastal eucalypt woodland over annual* Sorghum. **Bottom**—*inland eucalypt and* Terminalia *open woodland over white grass*

Figure 2.6 Tree density

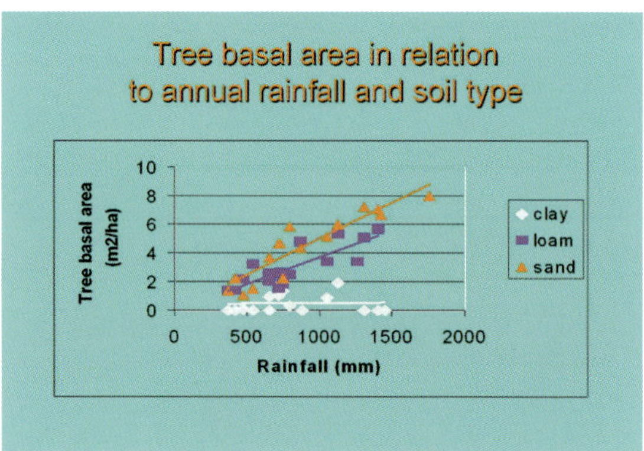

Trees are denser under higher rainfall and on lighter soils (sands) than on loams and clays.

Figure 2.7 Plant species richness

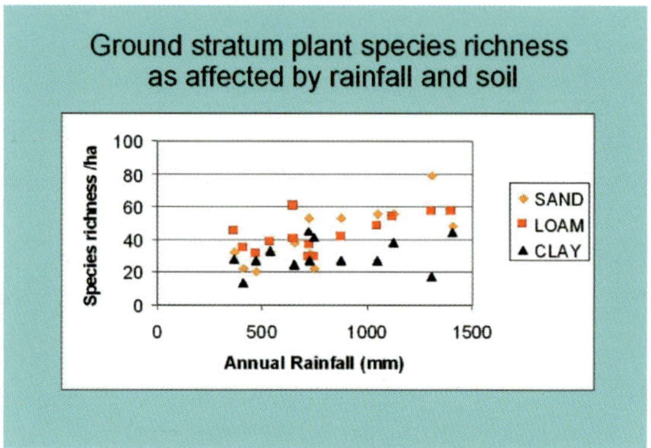

Biodiversity—increases with increasing rainfall and is generally greater on lighter (sandier) soils than heavier (clay) soils.

As annual rainfall increases, Mitchell grasses are replaced by blue grasses (*Dichanthium* spp.) and, in some parts of north-western Australia, by annual sorghums.

Other true grasslands in northern Australia are far less extensive and include those on sub-coastal flood plains (dominated by *Hymenachne*, *Oryza* and *Xerochloa* spp.) and littoral zones (dominated by *Sporobolus virginicus*). Other grasslands, such as blady grass (*Imperata cylindrica*), may result from tree clearing, often in conjunction with sown, introduced pasture species.

Spinifex may form true grassland, but is more frequently a ground layer under scattered trees dominated by acacias. These occur on very large areas of generally sandy soils in the driest savannas of northern Australia, such as parts of the Tanami Desert.

Structure and diversity—the importance of annual rainfall and soil type

Tree height, cover, density and basal area (the total cross-sectional area of all tree trunks) increase predictably with increasing rainfall and declining soil clay content. The pattern for the Top End of the NT is illustrated in Figure 2.6.

The diversity of plants and animals and the mix of species in a given area are also related to annual rainfall and soil clay content. Figure 2.7 illustrates such a pattern for plants in the NT. Tree cover, the cover of tall annual grasses, and the diversity of plants and some vertebrates increase with increasing annual rainfall, and are generally lower on the clay soils than on the sands and loams.

The vertebrates of the woodlands on the lighter–medium textured soils tend to be more diverse than those of the more nutrient-rich clay soils. The composition of mammal populations, in both space and time, may vary with long-term changes in the distribution of permanent water. The mix of invertebrates may also vary with annual rainfall and soil type, but are also strongly affected by the seasonal patterns of water.

Seasonal patterns

Growth and reproduction in both plants and animals are strongly linked to the seasons. In the ground layer, growth, flowering and setting of seed in herbaceous plants occurs during the wet season. With the onset of the dry season most of these plants begin to die off or become dormant. However, the trees behave in a different manner. Most species flower and fruit and produce their main leaf flush

Figure 2.8 Seasonality of flowering

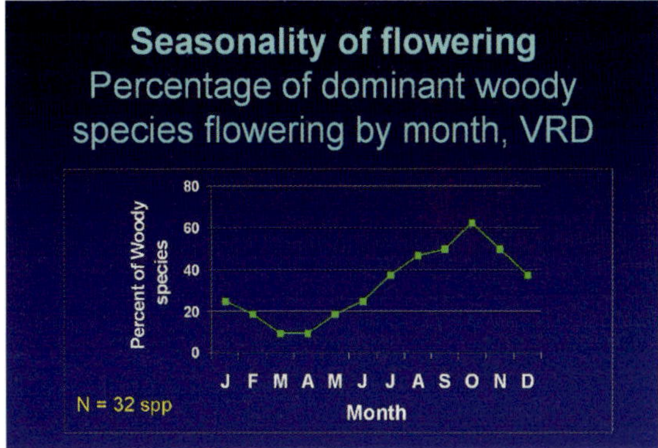

The dominant trees tend to flower during the dry season with a peak in the late dry season. Monthly pattern for 32 tree species in the Victoria River District (VRD)

in the dry season, usually peaking late in the dry season (Figure 2.8). This is an important food resource for animals during the dry season.

Many tree species hold seed-bearing fruit in the canopy for several years; seed is dispersed after fire. However, most savanna tree species hold their fruit for only a few months; seed is dispersed by the early wet season, and thus is ready to germinate with the first rains.

The animals of the savanna also show seasonal patterns. Invertebrates may have distinctly different species mixes during the wet and dry seasons. Vertebrates reproduce primarily in the wet season, but some do so in the dry season. There is some dry season dormancy, but this is limited to certain vertebrate groups (e.g. frogs and lizards).

Conclusion

Although the savannas of Australia are diverse, they all share a wet/dry climate and broad vegetation structure. These fundamental landscape characteristics make the savannas prone to frequent fire. The savannas are biologically 'active' throughout the year, including the dry season—the period when most fires occur. The interactions between fire and both the structure and seasonal patterns of growth and reproduction of the plants and animals are important factors that affect the impacts—both positive and negative—of fires in the savanna. The fires, their impacts, and the ways we can manage these interactions will be explored in detail in subsequent chapters.

Further reading

General

Finlayson, C. M. and von Oertzen, I. (1996). Eds *Landscape and Vegetation Ecology of the Kakadu Region, Northern Australia*. Kluwer, Dordrecht, The Netherlands.

Gillison, A. N. (1994). Woodlands. In *Australian Vegetation*. 2nd Edition. (Ed. R. H. Groves.) Cambridge University Press, Cambridge. pp. 227–255.

Mott, J. J., Williams, J., Andrew, M. A. and Gillison, A. N. (1985). Australian savanna ecosystems. In *Ecology and management of the World's Savannas*. (Eds J.C. Tothill and J. J. Mott.) Australian Academy of Science, Canberra. pp. 56–82.

Climate

Braithwaite, R. W. and Estbergs, J. (1988). Tuning in to the six seasons of the wet-dry tropics. *Australian Natural History* **22**: 445–449.

Cook, G. D. (1998). Rainfall reckoning in rangelands: The case of north Australian savannas. *Range Management Newsletter* **98**: 10–12.

McDonald, N. S. and McAlpine, J. (1991). Floods and drought: The northern climate. In *Monsoonal Australia. Landscape, ecology and man in the northern lowlands*. (Eds C. D. Haynes, M. G. Ridpath and M. A. J. Williams.) Balkema, Rotterdam, The Netherlands. pp. 19–29.

Taylor, J. A. and Tulloch, D. (1985). Rainfall in the wet-dry tropics: Extreme events at Darwin and similarities between years in the period 1870–1983. *Australian Journal of Ecology* **10**: 281–295.

Geology and geomorphology

Williams, M. A. J. (1991). Evolution of the landscape. In *Monsoonal Australia. Landscape, ecology and man in the northern lowlands*. (Eds C. D. Haynes, M. G. Ridpath and M. A. J. Williams.) Balkema, Rotterdam, The Netherlands. pp. 5–17.

Soils

Isbell, R. F. (1983). Kimberley-Arnhem-Cape York (III). In *Soils: An Australian Viewpoint*. (Ed. CSIRO Division of Soils.) CSIRO, Melbourne. pp. 189–199.

Isbell, R. F. and Hubble, R. F. (1983). North-eastern Plains. In *Soils: An Australian Viewpoint*. (Ed. CSIRO Division of Soils.) CSIRO, Melbourne. pp. 201–209.

Flora and fauna

Dunlop, C. R. and L. J. Webb. (1991). Flora and vegetation. In *Monsoonal Australia: landscape ecology and man in the northern lowlands*. (Eds C. D. Haynes, M. G. Ridpath and M. A. J. Williams.) Balkema, Rotterdam. pp. 41–61.

Fisher, A. (2000). Monitoring biodiversity in northern grassy landscapes. In *Proceedings of the Grassy Landscapes Conference* 2000. pp. 78–84.

Tothill, J.C. and Gillies, C. (1992). *The pasture lands of northern Australia: their condition, productivity and sustainability.* Occasional Publication No.5, Tropical Grassland Society of Australia, Brisbane.

Wilson, B. A., Brocklehurst, P. S., Clark M. J. and Dickinson, K. J. M. (1990). *Vegetation Survey of the Northern Territory, Australia.* Conservation Commission of the Northern Territory, Technical Report No. 49, Darwin.

Williams, R. J., Duff, G. A., Bowman, D. M. J. S. and Cook, G. D. (1996). Variation in the composition and structure of tropical savannas as a function of rainfall and soil texture along a large-scale climatic gradient in the Northern Territory, Australia. *Journal of Biogeography* **23**: 747–756.

Williams, R. J., Myers, B. A., Muller, W. A., Duff, G. A. and Eamus, D. (1997). Leaf phenology of woody species in a northern Australian tropical savanna. *Ecology* **78**: 2542–2558.

Williams, R. J., Myers, B. A., Duff, G. A. and Eamus D. (1999). Reproductive phenology of woody species in a north Australian tropical savanna. *Biotropica* **31**: 626–636.

Woinarski, J. C. Z. and Braithwaite, R. W. (1990). Conservation foci for Australian birds and mammals. *Search* **21**: 65–69.

Woinarski, J. C. Z., Fisher, A. and Milne, D. (1999a). Distribution patterns of vertebrates in relation to an extensive rainfall gradient and soil variation in the tropical savannas of the Northern Territory, Australia. *Journal of Tropical Ecology* **15**: 381–398.

3. Savanna fire regimes

by Dick Williams and Garry Cook

Fire regimes

Frequent and extensive fires in northern Australia are a consequence of the region's monsoonal climate with its marked summer wet season and long and warm winter dry season. The wet season generates heavy growth of grasses and other herbs, and the trees are continually dropping leaf litter throughout the dry season. This dries out or 'cures' during the dry season into tinder-dry, fine fuels for fires. Dry thunderstorms during the build-up and early wet season have always produced lightning.

Aboriginal people used fire widely across most of Australia and continue to do so across much of the north. Thus, for thousands of years there has been the combination of annual supplies of dry, fine fuel and ignition sources that can sustain regular, frequent fires.

European settlement has caused significant change to the fire patterns of northern Australia in the last century. The challenge for today's land managers is to work out the optimum mix of fire patterns at the landscape scale—hundreds to thousands of square kilometres—that protect life and property, maintain the productive potential of the land and conserve biodiversity. To do that, we need to understand the factors that determine the fire regimes in the savannas.

What are fire regimes?

The term 'fire regime' describes when and how often an area is burned, as well as the intensity, size and patchiness of fires (for example, yearly late dry season fires that are intense and extensive). Fire regimes can have different impacts on fuels, fodder and biodiversity by changing the composition of plant species and by altering habitats. Different fire regimes may be needed for different land management goals, and no single regime is best for all land management purposes.

How much savanna burns each year?

Hundreds of thousands of square kilometres of northern Australia are burnt each year. The area burnt varies with state or territory; in the northern savannas usually more country is burnt in the NT than in either northern WA or Queensland.

When do the fires occur?

Fires can occur from March to December, but most areas are burnt late in the dry season (Figures 1.1 and 3.1). Of the nine operational districts of the Bushfires Council of the NT for example, only one region east of Darwin has more country burnt in the early dry than the late dry season.

Fires can occur in the wet season, especially in the early part (November–December) if there is available fuel and if weather conditions allow fires to spread.

How often is an area burnt?

In the higher rainfall savannas of the NT, for example the Darwin–Alligator Rivers region, 50–70% of the landscape may be burnt every year (Figure 3.3). Fires are less frequent in the semi-arid savannas; in the mid-1990s, nearly 80% of the area south of Katherine (NT) was either unburnt or burnt only once over a three-year period.

What determines fire regimes?

Regional fire regimes depend upon the weather, the amount of fuel and when it is available, ignition chance and type, fire behaviour, and the capacity of fire to spread through the landscape. Fire regimes are also the product of people; people are the most common source of ignition and they can manipulate fire regimes, by prescribing fire or by suppressing it. People also affect fire regimes by their influence on fuel loads as a consequence of land use. For example, the difference in fire frequency between the semi-arid and mesic regions of the NT may be partly related to climate, but also to land use.

In the mesic regions, the dominant pasture type is annual *Sorghum*, a relively unpalatable grass, whereas in the semi-arid savannas, the pasture species are more palatable perennial grasses so that cattle may reduce fuel loads by eating the grass. Property managers burn less extensively for fear of losing their feed reserves.

In the following sections we will review the factors that affect fire regimes.

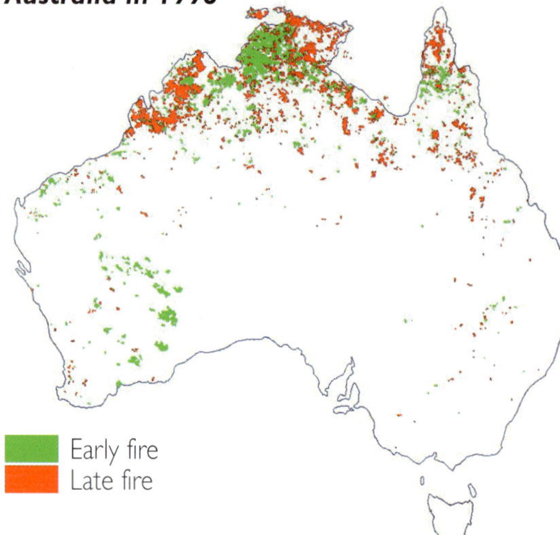

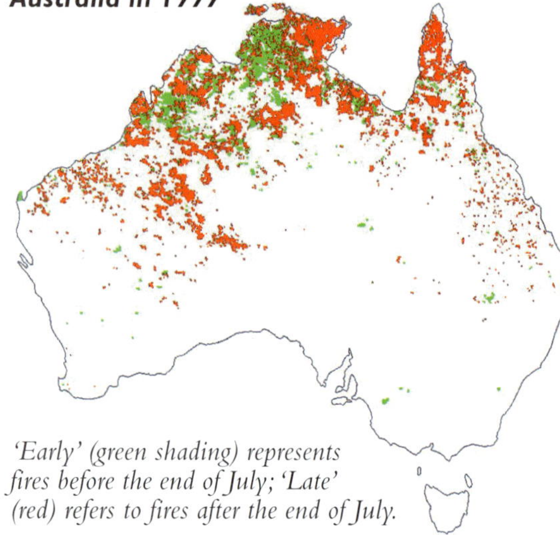

'Early' (green shading) represents fires before the end of July; 'Late' (red) refers to fires after the end of July.

Figure 3.3 Fire frequency

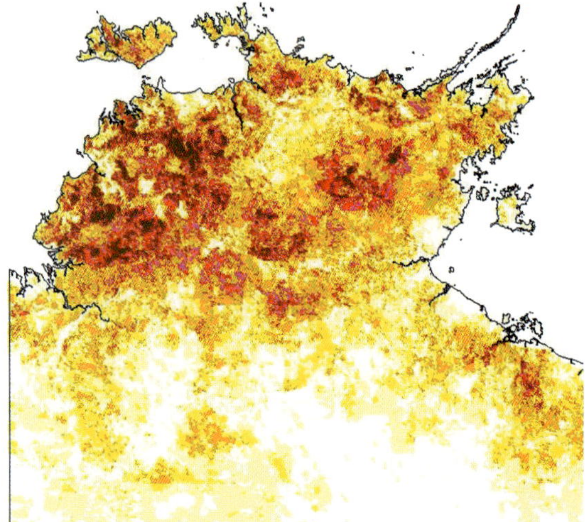

Frequency of fires (number of times burnt) in the Top End of the NT, 1993–2000

Times Burnt

Not burnt	3 times	6 times
Once	4 times	7 times
Twice	5 times	8 times

Fire weather and fire danger

Fire weather is a term used to describe the prevailing weather conditions as they relate to fire behaviour. The primary weather variables that determine fire behaviour are atmospheric humidity, air temperature and the strength of the winds. Variation in these factors in northern Australia is governed by the arrival and departure of the monsoon, hence there is strong seasonal variation in fire weather.

Fire weather variables, in conjunction with fuel variables, can be incorporated into indices that express fire danger. Fire danger is the product of the factors that determine chance of ignition, propensity to spread and ease of suppression. Summaries of fire weather and fire danger are given in the books by Luke and McArthur, and Cheney and Sullivan (see Further reading).

The two main fire danger indices used in Australia were developed by A. G. McArthur in the 1960s— the Forest Fire Danger Index (FFDI) and the Grassland Fire Danger Index (GFDI). FFDI includes a drought factor while GFDI includes a fuel-curing factor. Fire danger rating systems allow public fire weather forecasts to be made by the Bureau of Meteorology, and provide fire management agencies (or other land managers) with up-to-date, reliable information for purposes of fire control or suppression.

Both FFDI and GFDI rise as temperatures and wind speed increase and as relative humidity and soil moisture drop. The indices range from 0 to 100, with values above 50 considered extreme. Luke and McArthur indicate that an index of 100 represents the 'near worst possible fire weather conditions that are likely to be experienced in Australia', although Cheney and Sullivan indicate that such values have been exceeded on several occasions since 1966. A value of 100 would be produced by air temperatures of 40°C, relative humidities of 15%, wind speeds of 55 km/h, an extended period of drought of 6–8 weeks or more and abundant fuel. Fires that burn under extreme conditions are virtually impossible to put out without massive effort.

Both indices can be calculated easily using fire danger meters. In the savannas of northern Australia, fire danger is calculated using a modified Grassland Fire Danger Meter (Figure 3.4c) because savanna fires, even in woodlands, behave more like grass fires than forest fires. There are five operational classes of fire danger: Low (<2.5); Moderate (2.5–7.5); High (7.5–20); Very High (20–50) and Extreme (>50). Calculations are made by the Bureau of Meteorology with appropriate operational responses taken by the responsible fire management agencies (both rural and urban) and other land managers.

The indices may be used to compare seasonal variation in fire weather but there have been few systematic studies of this in the savannas. One such study, from Jabiru in the NT, examined seasonal variation in FFDI and GFDI based on 12 years of records.

The patterns of afternoon (3 p.m.) FFDI show strong seasonality (Figures 3.4a, 3.4b). The average daily FFDI is below five during the peak monsoon period of January to early March, and the vegetation will generally not burn. The wet season ends abruptly, and both atmospheric and soil moisture drop from late March onwards. Despite this, average daily FFDI in the early dry season (May–June) remains below 20, steadily increasing to average daily values of around 20 in the September–October period. However, the average maximum FFDI during this period of peak fire weather averages about 40— Very High. The highest value recorded was 60, and values >40 occurred in all months from June– October. These extreme values, despite being in the High–Very High classes, are below peak levels of 100 that can occur on extreme days in southern Australia, such as on Ash Wednesday in 1983.

The FFDI declines again with the onset of the wet season, but the maximum can still be around 20–30 on some days in November–December. Under such conditions, fire spread is possible, allowing wet season burning to be undertaken.

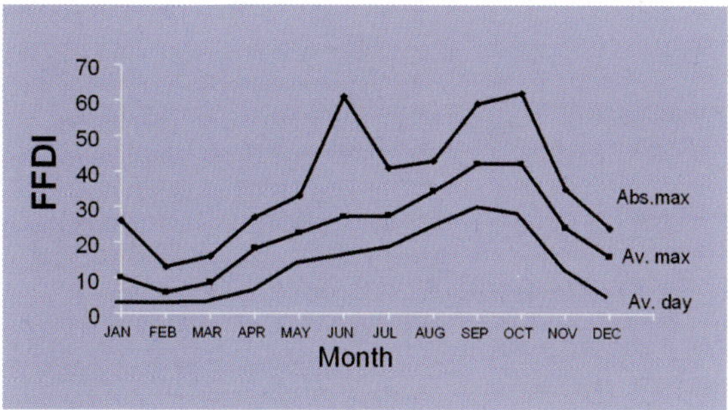

Figure 3.4a Forest Fire Danger Index, Jabiru

Forest Fire Danger Index increases as the dry season progresses, as temperatures and wind speed rise and relative humidity drops. Indicies based on 3 p.m. weather data from 12 years' records at Jabiru, NT; average annual rainfall 1300 mm. Source: Gill et al. (1996)
Abs. max = absolute maximum FFDI
Av. max = average monthly maximum FFDI
Av. day = average daily FFDI

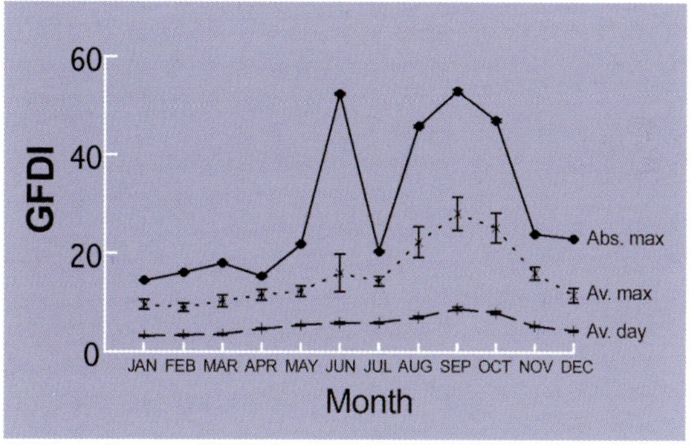

Figure 3.4b Grassland Fire Danger Index, Jabiru

Grassland Fire Danger Index for 100% cured grass fuels. GFDI also increases as the dry season progresses, as temperatures and wind speed rise and relative humidity drops. Data and legend as in Figure 3.4a

Figure 3.4c Grassland Fire Danger Meter

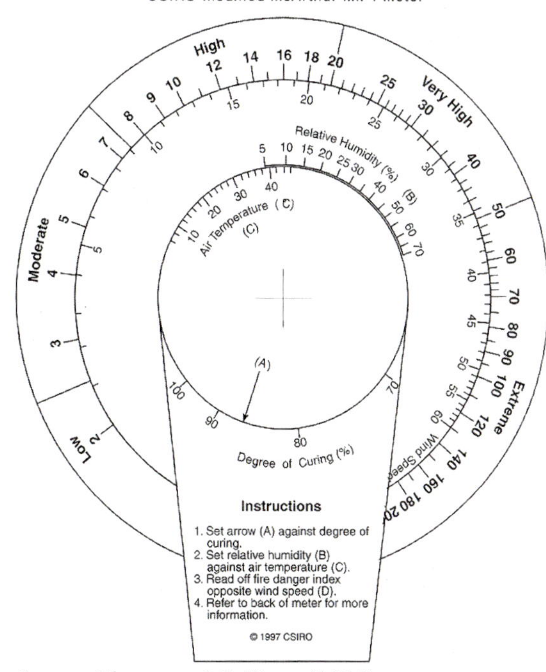

Source: Cheney and Sullivan (1997)

Fuel dynamics

There is no fire without fuel. The amount of fuel—its height, mass, composition and architecture—all vary annually and seasonally, and with land use. Fuel load is an important determinant of fire intensity but, unlike the weather, it is something the landholder can manipulate.

What burns?

Most savanna fuels are fine fuels, less than 6 mm in diameter in at least one dimension. Grass provides most of the fine fuel although trees can drop increasing amounts of leaf and twig as the dry season progresses. The amount of tree leaf litter in the fuel bed increases with increasing tree density, which in turn rises with increasing annual rainfall. In higher rainfall areas, tall annual grasses such as *Sorghum* create fuel loads that dry quickly after flowering.

The ribbons of bark and layers of shrubs that create 'ladder' fuels in the eucalypt forests of southern Australia rarely occur in northern Australia. Thus the flames in savanna fires generally remain within 5 m of the ground, and severe 'crown fires' in tree canopies rarely occur.

Fuel dries out as the dry season progresses. The tall annual grasses start drying during the late wet season (March–April) while perennials can last until the early dry season (June).

In the semi-arid savannas where the ground layer is dominated by perennial grasses, fine grassy fuels may accumulate to levels similar to those in the wetter savannas. Tall annual grasses are rare in these savannas; where annuals occur they are usually short and sparse and often preferentially grazed. They usually have little bulk, especially in the dry season and, if burned, would support only patchy low-intensity fires.

Termites are abundant on the less fertile soils of the north, and their total body weight per unit area may be greater than that of large herbivores. They remove considerable grass stem and litter during the dry season.

Fuel loads fluctuate through the years and seasons, and can be modified by grazing. Grass builds up rapidly during the wet season, but in pastoral lands some may be eaten by stock. In practice, pasture land may have to be rested, with stock totally removed for one good wet season to allow enough fuel to accumulate so that fire can occur.

Fuel cover over the ground needs to be relatively unbroken with more than 50% ground cover needed to support a continuous fire front.

How much fuel accumulates?

Fuel loads need to be understood to assess the risk of wildfire or for planning prescribed fires. The fuel load is the oven-dry mass of fuel in a given area.

Dry grass provides the bulk of the fuel. A continuous fire front needs at least 50% ground cover.

Tree leaf litter can be a significant component of the fuel. It increases as the dry season progresses and with increasing tree density.

Termites gather stems of Sorghum *and wiregrass.*

Fuel loads can be expressed in different units—tonnes per hectare, kilograms per hectare, kilograms or grams per square metre.

Annual inputs of fine fuel from grass growth and litter fall are about 2–8 tonnes of dry matter per hectare. Under annual burning, fuel loads generally remain around these levels in both higher rainfall and semi-arid savannas. If country remains unburnt, the fuel loads rise but do not generally increase above the equivalent of a few years' grass growth and leaf fall because the litter breaks down quickly during each wet season.

Various studies have measured how fuel loads vary in different types of savanna country.

In the wetter regions, fuel loads can reach an equilibrium (between accumulation and decomposition) of about 10 t/ha, 2–3 years after fire.

Perennial grass pastures near Katherine, NT (950 mm rainfall) had grass fuel levels of 2–4 t/ha when burnt every two years and reached 6 t/ha when protected from fire for four years.

In Rockhampton, central Queensland (890 mm rainfall), loads reached 3 t/ha in annually burnt savannas, and 6-7 t/ha after three years without fire.

In sparse woodland with spinifex, where fires are less frequent, fuel loads can reach 10–20 t/ha after 5–10 years. However, in the spinifex-dominated landscapes of the stone country in the wetter savannas, for example the Arnhem Land Plateau, fuel loads may reach such levels in less than five years if unburnt.

Where exotic grasses invade and are not eaten, fuel loads can increase dramatically. With some introduced species such as gamba grass, fuel loads may be 20 t/ha, 4–5 times the normal fuel load for a savanna.

What is fuel curing?

Curing refers to the 'greenness' of the fuel; it indicates the moisture content of the fuel, its flammability and thus the potential rate of fire spread. Curing starts as soon as a grass has flowered and its stem becomes brittle; this is greatly accelerated by the sudden onset of the dry season.

Fuel curing can vary greatly across the landscape according to position in the landscape, the dominant grasses and the season. Effective fire management over broad areas has to take this into account. For example, grasses such as annual *Sorghum* on sandy soils will cure rapidly and therefore be flammable much earlier than perennial grasses on cracking clays. In general, by the end of the dry season, the fine grassy fuels of the savannas are fully cured.

Satellite imagery can help identify areas in different states of curing across the landscape. Hence we can identify those areas at greatest risk from uncontrolled fire, or those that are ready for prescribed fuel reduction burning and how prescribed fires are likely to be spread.

One way to 'view' such features of the landscape is by using NDVI (Normalised Difference Vegetation Index), derived from satellite imagery. Maps of NDVI show areas of different chlorophyll reflectance, indicating fuel 'greenness', and therefore how cured the fuel is. Photographic standards can be used to estimate the degree of fuel curing (see Chapter 7. Monitoring vegetation greenness).

Grass fuel loads accumulate with time, unless eaten by animals or until the old leaf decomposes during the wet season. In the savannas, fine fuels tend not to accumulate above 10 t/ha. Light grazing (left) and heavy grazing (right)

Photo standards illustrating fuel loads 1.6–4.5 t/ha, in the VRD region, NT

1.6 t/ha in ribbon grass/Flinders grass

2.4 t/ha in black speargrass

3.5 t/ha in white grass

4.5 t/ha in bluegrass

Photo standards illustrating green, uncured fuel (in March) and fully cured fuel (in October) on red earths and black cracking clays, VRD region, NT

Fuel on red soil, March
1.9 t/ha; 35% cured; 40% moisture content

Fuel on red soil, October
2.2 t/ha; 95% cured; 10% moisture content

Fuel on black soil, March
1.3 t/ha; 35% cured; 55% moisture content

Fuel on black soil, October
1.5 t/ha; 90% cured; 15% moisture content

Exotic grasses and fire

Gamba grass (*Andropogon gayanus*) and mission grass (*Pennisetum polystachion*) are African perennial species used for pasture trials at Katherine in the 1940s and '50s. Gamba grass has been successful as a pasture species, but both species have become serious environmental weeds in the Darwin region and parts of Cape York when not managed.

On the rural–urban interface of Darwin where development and disturbance coincide, these grasses grow and spread prolifically. The establishment of gamba grass on flood plain margins and wetter *Melaleuca* uplands is enhanced by soil disturbance while in *Eucalyptus* woodlands disturbance is not essential. Establishment of gamba grass is significantly higher in *Eucalyptus* woodland that has recently been burnt.

These introduced grasses produce flammable material up to five times greater than fuel loads in native grasses (which are typically 2–8 t/ha), cure later in the year (June–July versus April) and maintain a tall, upright structure. Fire intensities and flame heights are increased and have severe impacts on the less fire-tolerant native grasses, shrubs and trees. Recovery from fire by these exotic perennials is rapid.

Gamba grass has a relatively short-lived seed bank with only 1% of buried seed remaining viable within a year; thus gamba grass infestations can be eradicated over a couple of years. Glyphosate herbicide is effective on gamba grass and mission grass especially if fire can be used to remove rank growth first.

by Trevor Howard

Further reading

Flores, T. (1999). Factors affecting the recruitment of *Andropogon gayanus* Kunth (Gamba Grass), Honours Thesis, Northern Territory University.

Panton, W. (1993). Changes in post World War 11 distribution and status of monsoon rainforests in the Darwin area, *Australian Geographer* **24(2):** 50–59.

Gamba grass on roadside, Darwin region, NT

Mission grass on roadside, Darwin region, NT

Trees burnt by fire in gamba grass, NT

Ignition line in gamba grass, NT

Fire behaviour

Fire behaviour describes the physical attributes of individual fires—the height and depth of the flames, the speed with which the fire moves, the size and shape of the various fronts, and the intensity of the fire. Many of these attributes can be inferred from post-fire features of the burnt landscape, such as the height of blackened or scorched leaves, the size of standing twigs consumed by the flames, and the degree of fuel consumption, i.e. fire patchiness. Knowledge about fire behaviour is important to understand the likely extent and effectiveness of fires—be they wanted or unwanted—that may affect our patch of land, and so that we can understand the effects of fires on the landscape and atmosphere.

What is fire intensity?

One measure of fire behaviour is 'fire intensity'. This represents the rate at which energy is released, and is measured as kilowatts (a unit of work) per metre of fire front. It is a function of the heat yield of the fuel (heat per unit mass burnt as kilojoules/kg), the amount of fuel per unit area (kg/m^2) and the rate of forward spread of the fire front (m/second).

These three components of fire intensity can be measured or estimated.

Fuel loads can be determined by direct harvest, by using indirect pasture-estimation techniques such as BOTANAL or by using calibrated, photo-graphic fuel standards (see p. 20). It is important to estimate the degree of fuel consumption, i.e. the proportion of the available fuel that will be burnt.

Heat yield is often assumed to be 20,000 kJ/kg for mixed fuel types. However, the heat content of various fuel components varies between species.

Rate of spread is the hardest component to measure, but it can be estimated if the time to arrival of the flame front at three or more points can be determined accurately (within one second). The most important determinants of rate of spread are wind speed (see Fire Behaviour p. 24), relative humidity and fuel moisture. These factors vary throughout the day and seasonally, from early dry season to late dry season. Rate of spread—and hence intensity—can therefore be manipulated by careful consideration of ignition time—both during the day (or night) and from month to month.

Savanna fires generally move at speeds of 0.1–2 m/s. Fuels loads are generally in the range of 2–8 t/ha, with fine fuel consumption rates of 50–100%. Fire intensities in general range from 500 to 10,000 kW/m, and rarely exceed 20,000 kW/m. In southern Australian eucalypt forests where fuel has accumulated to near maximum levels (in excess of 30 t/ha), fire intensities can be as high as 50,000–100,000 kW/m.

Over a five-year period at Kapalga in Kakadu National Park, early dry season fires (lit in early June) averaged about 2000 kW/m whereas late dry season fires (lit in late September) averaged about 8000 kW/m.

Flames from 500–1000 kW/m fires are less than 1 m high, but can reach 2–4 m if the intensity is above 5000 kW/m.

A visual estimation of fire intensity is illustrated on page 25.

An indication of flame height is given by char height—the height above ground of blackened leaves that are still attached to trees or shrubs. Char heights increase by about a metre for every 2500 kW/m (Figure 3.5a).

Figures 3.5a and 3.5b Char and scorch height

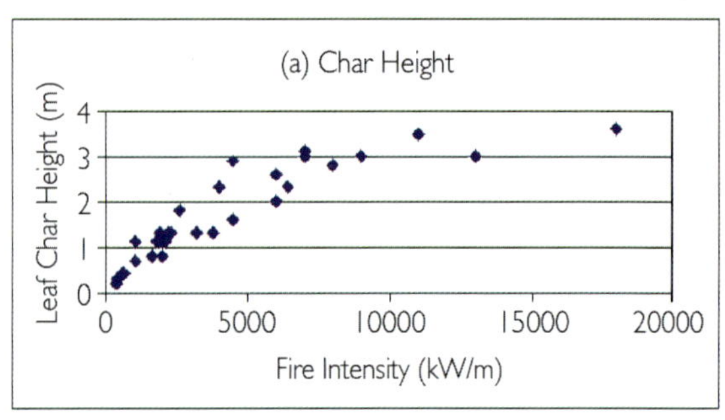

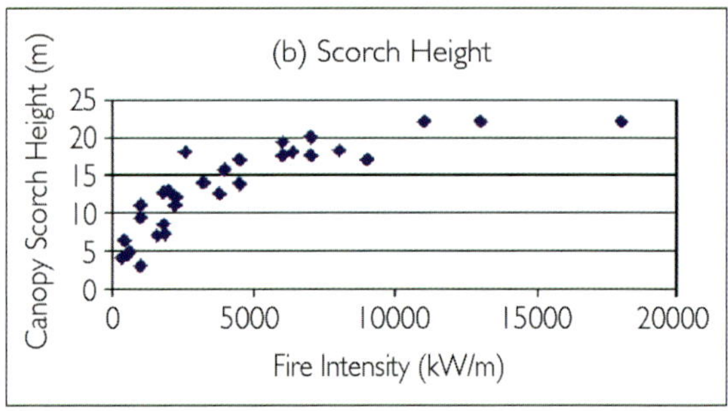

The height of charring (blackened leaves) and scorching (brown leaves) on woody plants can be used to gauge the intensity of a fire.

Leaf scorch height—the height above ground to which the leaves in the tree canopy are killed, and thus 'browned'—also varies with intensity. Scorch height increases by about 3 m for every 1000 kW/m of fire intensity up to about 8000 kW/m (Figure 3.5b). At higher intensities the tops of the canopies of even the tallest trees are scorched.

Fires greater than 2000 kW/m tend to burn all the available fuel whereas less intense fires create a mosaic of burnt and unburnt patches across the landscape.

These rules of thumb concerning crown and ground scorch can be used to gauge the intensity of a fire after it has passed.

Ignition type is an important factor affecting fire behaviour. Point sources of ignition lead to elliptical-shaped fires (Figure 3.6). Only the country at the head of the ellipse is burnt with maximum intensity; the flanking country on the sides and back end of the ellipse is burnt at much lower intensities. Thus there is considerable variation in fire intensity across the landscape.

Line ignitions, such as those along roadsides, lead to fires burning on a broad front; such fires accelerate to maximum rates of spread very quickly—in minutes. Thus, compared with a point-source ignition, fire intensity is more uniform across the front, more of the country is affected by the heading fire, and there is less variation in intensity across the landscape. Perimeter fires—lighting up lines on more than one front—can also create broad, fast-moving, more intense fires. These variations in behaviour, as a consequence of ignition type, have implications on how we may use fire and why, as is illustrated in more detail in Chapter 6.

Figure 3.6 Elliptical-shaped fires

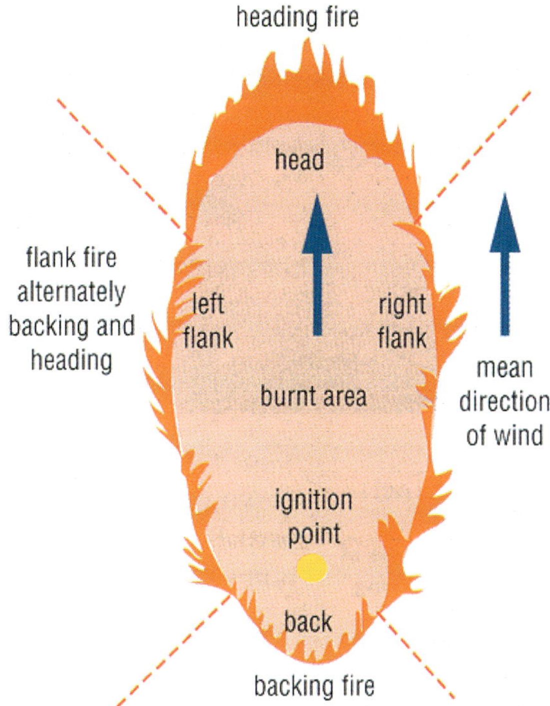

Fires that start from point sources develop an elliptical shape. The ellipse consists of a heading fire, flanking fires and a backing fire, all of which vary in flame and combustion characteristics. Source: Cheney and Sullivan (1997)

Elliptical-shaped fires

Fire behaviour

Flame characteristics

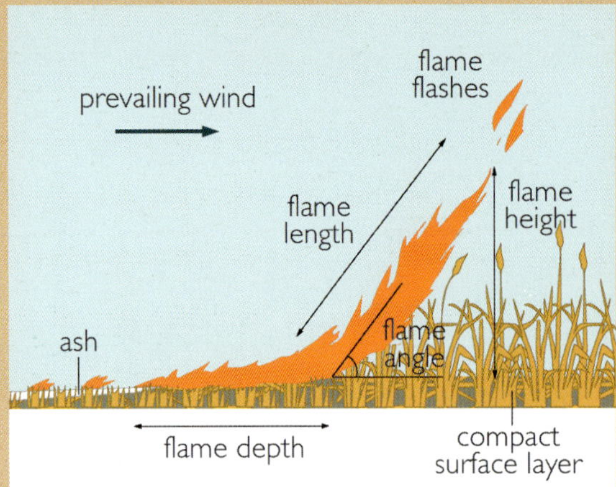

The height of flames depends upon the degree of combustion of standing grass; depth of flame depends on the amount of fuel in the compacted surface layer. Source: Cheney and Sullivan (1997)

The rate of forward spread

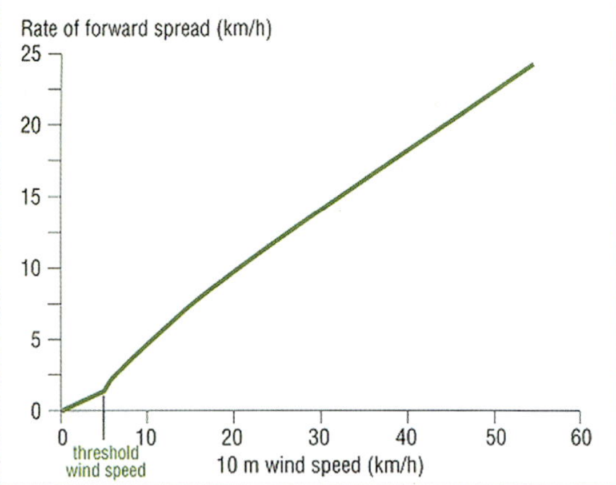

The rate of spread of a fire depends on wind speed. Source: Cheney and Sullivan (1997)

A heading fire (top) and a backing fire (bottom)

Estimating fire intensity

A quick estimate of fire intensity can be obtained from the equation

Intensity = 500 x W x R

where W is the fuel consumed in t/ha and R is spread in km/h

Thus an annual *Sorghum* grassland with 2 t/ha spreading at 5 km/h will have a fire intensity of

Intensity = 500 x 2 x 5 = 5000 kW/m

Low-intensity fire (1.4 t/ha; 2700 kW/m fuel—left) and moderate intensity fire (6.7 t/ha; 5200 kW/m fuel—right) in woodland with perennial grass understorey at Kidman Springs, Victoria River District, NT

High-intensity fire (20,000 kW/m) in open forest at Kapalga, NT

Fire regimes and landscape management

Traditional Aboriginal burning

Indigenous people have traditionally used fire for a variety of reasons—ease of travel, communication, 'cleaning' country, hunting and to maintain food sources. In Kakadu, for example, they started burning very late in the wet season (March) as soon as the country began to dry out, and continued for 9–10 months until the monsoon arrived. There was a peak of activity in June–July, but relatively little in August–September.

Current fire regimes on Aboriginal lands may not reflect past traditional practices because people are no longer dispersed through the country, and because traditional knowledge may not have been passed on to the current generation. Fires arising from Aboriginal lands today can cause conflict with neighbouring landholders if their land management goals are different.

It is important to continue learning about traditional indigenous fire regimes. Today, fire agencies endeavour to employ traditional burning practices, where possible, as a means of establishing a diverse set of fire regimes across the savanna landscape.

Early dry season burning

Early dry season burning is often prescribed in all land tenures in the savannas, primarily to reduce fuel. The early dry season burn results in a mixture of burnt and unburnt country and the risk of extensive, intense fire in the later part of the dry season is reduced. Early dry season fire can also be used to stimulate 'green pick' in pastoral country.

Late dry season fires

Most of the area burnt in northern Australia results from late dry season fires—wildfires rather than planned and controlled operations. Most late dry season fires are undesirable as they are difficult to suppress, consume vast tracts of country, and can be detrimental to plant and animal life.

However, intense, late dry season fires do have some uses. They can be used to kill or reduce the amount of woody sucker or shrub growth that may have accumulated in unburnt or heavily grazed savanna country. They can be used in integrated programs to control exotic woody weeds such as rubber vine (*Cryptostegia grandiflora*) in Queensland.

Wet season burning

The opportunity for a burn during the wet season depends on the chance of a dry break during the monsoon. For wet season burning, a body of cured grass and fuel needs to be retained through the dry season and then lit during the wet season after a couple of

days without rain. Burning soon after the first storms, or when storms are imminent, removes old rank grass just before the new growth of the next season and so improves the quality of fodder for livestock or native fauna.

An increasingly important use of wet season burning in the Top End of the NT is to manage stands of annual sorghum. If annual sorghum is burnt after germination but before seed set, it can be eliminated locally because it does not have a persistent seed bank. This can reduce fuel loads in country with sorghum by 50%. However, the effectiveness of wet season burning depends very much on timing. If too early in the wet season, before about 100 mm of rain has fallen, there may be ungerminated seed left in the soil. If too late, e.g. after the monsoon has set in, fire may not spread.

Early wet season fire or 'storm burns' can also be used to control the density of *Melaleuca* shoots in the grassy wetland country on Cape York and to manipulate the composition of the grasses in the grasslands. This is an important tool in the management of habitats for species such as the golden-shouldered parrot.

Fire regimes—past, present and future

There has been considerable change to the fire regimes in the savannas over evolutionary, pre-historic, historic and contemporary times.

Fires undoubtedly became more frequent as the climate began to dry out during the Tertiary period over 20 million years ago. Before Aboriginal people arrived in the savannas, fires probably started at the very end of the dry season when there was dry fuel and the early storms produced lightning.

Aboriginal people modified the timing of fire by burning earlier in the dry season. Now, much of northern Australia, particularly the wetter parts, is dominated by a regime of frequent, extensive fires in the late dry season (Figure 1.1 p. 1). These extensive, late dry season fires appear to be more common now than before European settlement in northern Australia. This regime of frequent, intense, extensive fires is cause for concern in virtually all savannas.

In much of the semi-arid country, the dominant land use is pastoralism for which reserves of grass need to be protected from fire. Fires have been actively suppressed, which has probably resulted in a decrease in fire frequency and extent.

A change in fire regimes brings with it a change in impacts of fire on the landscape. We therefore need to know the ways in which various regimes impact upon the plants and animals of the savannas, and how we can manipulate fire regimes so that desired impacts are maximised and undesirable ones are minimised. The following chapters provide more detail on the impacts of various regimes and how we can manipulate fire regimes to achieve the types of savanna landscape we want.

Further reading

General

Cheney, N. P. and Sullivan, A. (1997). *Grassfires: Fuel, weather and fire behaviour.* CSIRO, Melbourne.

Gill, A. M., Hoare, J. R. L. and Cheney, N. P. (1990). Fires and their effects in the wet-dry tropics of Australia. In *Fire in the Tropical Biota. Ecosystem Processes and Global Challenges.* (Ed. J. G. Goldammer.) Springer-Verlag, Berlin. pp. 159–178.

Fire weather

Gill, A. M., Moore, P. H. R. and Williams, R. J. (1996). Fire weather in the wet-dry tropics: Kakadu National Park, Australia. *Australian Journal of Ecology* **21:** 302–308.

Luke, R. H. and McArthur, A. G. (1978). *Bushfires in Australia.* Australian Government Publishing Service, Canberra.

Noble, I. R., Bary, G. A. V. and Gill, A. M. (1980). McArthur's fire danger meters expressed as equations. *Australian Journal of Ecology* **5:** 201–203.

Tapper, N. J., Garden, G., Gill, J. and Fernon, J. (1993). The climatology and meteorology of high fire danger in the Northern Territory. *Rangeland Journal* **15:** 339–351.

Fuels

Cook, G. D. (1994). The fate of nutrients during fires in a tropical savanna. *Australian Journal of Ecology* **19:** 359–365.

Fire behaviour

Cheney, N. P., Gould, J. S. and Catchpole, W. R. (1993). The influence of fuel, weather and fire shape variables on fire-spread in grasslands. *International Journal of Wildland Fire* **3:** 31–44.

Williams, R. J., Gill, A. M. and Moore, P. H. R. (1998). Seasonal changes in fire behaviour in a tropical savanna in northern Australia. *International Journal of Wildland Fire* **8:** 227–239.

Fire regimes

Braithwaite, R. W. (1991). Aboriginal fire regimes of monsoonal Australia in the 19th century. *Search* **22:** 247–249.

Craig, A. (1997). A review of information on the effects of fire in relation to the management of rangelands in the Kimberley high-rainfall zone. *Tropical Grasslands* **31:** 161–187.

Crowley, G. M. and Garnett, S. T. (1999). Changing Fire Management in the Pastoral Lands of Cape York Peninsula: 1623 to 1996. *Australian Geographical Studies* **38:** 10–26.

Fensham, R. J. (1997). Aboriginal fire regimes in Queensland, Australia: Analysis of the explorers' record. *Journal of Biogeography* **24:** 11–22.

Gill, A. M., Ryan, P. G., Moore, P. H. R. and Gibson, M. (2000). Fire regimes of World Heritage Kakadu National Park, Australia. *Austral Ecology* **25:** 616–625.

Haynes, C. D. (1985). The pattern and ecology of munwag: traditional Aboriginal fire regimes in north-central Arnhem Land. *Proceedings of the Ecological Society of Australia* **13:** 203–214.

Press, A. J. (1988). Comparisons of the extent of fire in different land management systems in the Top End of the Northern Territory. *Proceedings of the Ecological Society of Australia* **15:** 167–175.

Russell-Smith, J., Lucas, D., Gapindi, M., Gunbunuka, B., Kaparigia, N., Namingum, G., Lucas, K., Giulani, P. and Chaloupka, G. (1997). Aboriginal resource utilisation and fire management practice in western Arnhem Land, monsoonal northern Australia: notes for pre-history, lessons for the future. *Human Ecology* **25:** 159–195.

Russell-Smith, J., Ryan, P. G. and Durieu, R. (1997). A Landsat MSS-derived fire history of Kakadu National Park, monsoonal northern Australia, 1980–94: seasonal extent, frequency and patchiness. *Journal of Applied Ecology* **34:** 748–766.

Williams, R. J., Griffiths, A. D. and Allan, G. E. (2001). Fire regimes and biodiversity in the savannas of northern Australia. In *Flammable Australia: the fire regimes and biodiversity of a continent.* (Eds Bradstock, R. A., Williams, J. E. and Gill, A.M.) Cambridge University Press, Cambridge. pp. 281–304.

4. Effects of fire in the landscape

by Jeremy Russell-Smith, Tony Start and John Woinarski

Fire affects all aspects of the ecology of the savanna—individual plants, plant communities, animals and their habitats, nutrients, water catchments and down-stream hydrology. In turn, fire regimes have various effects ranging from major short-term impacts associated with frequent high-intensity fires, through to slower, longer-term changes associated with total fire exclusion.

In order to maintain savanna species diversity at landscape scales, an ideal fire regime requires the development of a fine-scale mosaic where burnt patches represent a range of fire intensities and fire ages—from recently burnt through to long unburnt. All land managers need to understand the ecological consequences of fire management.

Fire effects on plants

Natural selection in the fire-prone environments of tropical savannas has produced many plants that can tolerate defoliation and even the destruction of their stems by fire. The mechanisms used to survive include protected growing points, dormant buds and regularly replenished seed banks. Different plant groups tend to exploit strategies that suit their life histories and growth forms.

To examine strategies used by the different groups it is convenient to classify them as annuals (species that produce a new generation each year) and perennials (species that often live for several years). Both groups contain grasses and broad-leaved plants. Broad-leaved annuals are often called forbs while broad-leaved perennials include many growth forms but the most prominent in savannas are shrubs and trees.

How do annuals survive?

Annuals regenerate from seed each year, germinating when moisture conditions are suitable. Growth and reproduction are rapid and by the early dry season there is a fresh crop of seed on the soil. The spent plants are dying anyway so burning after seed set does not affect survival of the species.

Fires in the wet season or very early dry season may intercept seed production, but these fires are usually small, cool and patchy. The species survives because some escape damage in unburned areas from where they can quickly recolonise burnt patches. Seed of many species can remain dormant in the soil for more than one year.

How do perennial grasses survive?

Most perennial grasses form dense tussocks, sending up new shoots from bud sites at or below ground level. Fire may destroy the tiller leaves or stems (which are often dead anyway in the dry season), but the protected buds can quickly shoot.

Indeed 'green pick' can often be seen a week or two after fire. Individual tussocks will die at some stage so even perennial grasses depend on seed to perpetuate. Excessive grazing reduces seed set while heavy grazing and frequent intense fires may reduce tussock size. Annual species can then invade.

Some perennial grass plants, notably spinifex, may be killed by fire but regenerate from seed. Being relatively slow growing and well spread out, these grass hummocks do not amass sufficient fuel to carry another fire until they have replenished their seed store in the soil.

Annuals regenerate from seed each year.

Perennial grasses recover from tiller bud sites located at or below ground level.

How do trees and shrubs survive?

Some shrubs are killed by fire and have to regenerate from seed. However, the stems of most woody plants, particularly trees, are protected by thick layers of fibrous or corky bark that insulate the underlying live tissue. If the foliage and smaller branches are killed by fire, dormant buds under the bark of larger stems are activated and the foliage is replaced by new shoots (Table 4.1).

Many shrubs and young trees coppice from dormant buds at the base of the plant even if their stems are killed to ground level. Some species develop large woody structures—lignotubers—which are studded with dormant buds just below the ground surface.

Eventually, individuals in all these groups will die from one cause or another and so all need to recruit seedlings to persist in a community.

The most resilient trees and shrubs are those that can shoot from their roots. Hypothetically, they can live indefinitely, even if their stems are regularly destroyed, because they are continually growing new roots from which they can re-shoot long after the original parent stem has died and decomposed or burnt away.

How else can perennial plants survive?

Some plants—cycads, some palms, pandanus and even some perennial sundews—shelter a growing tip in the midst of a crown of fire-resistant leaf bases. Others—yams and lilies—survive the dry season (and fire) as underground rhizomes, starchy tubers or bulbs, sending up their aerial shoots during the wet season.

Table 4.1 Attributes of fire-tolerant and fire-sensitive woody plants

Attribute	Fire-tolerant	Fire-sensitive species
Resprouting	Yes	No
Must regenerate from seed after fire	No	Yes, but some species have limited ability to resprout
Critical time for regeneration after fire	No	Years needed after seed germination to replenish seed bank

Many shrubs and small trees will coppice from their bases.

Cycads regrow from crowns of fire-resistant leaf bases.

Savanna trees are protected from fire by thick or corky bark.

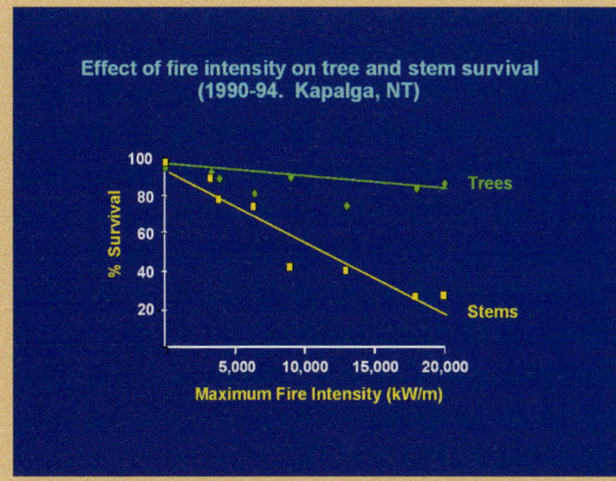

Effect of fire intensity on tree and stem survival (1990-94. Kapalga, NT)

How fire affects savanna trees

The responses of savanna trees to different fire regimes can be illustrated from findings at Kapalga in Kakadu National Park.

1   Increasing fire intensity increases death of tree stems (low-intensity fires are typically less than 2000 kW/m), but most individuals resprout from the base of stems.

Effect of fire regime on live tree basal area

2   Plots with early fires each year maintain the same woody biomass (as measured by live tree basal area) as unburnt plots. Late dry season fires each year reduce woody biomass.

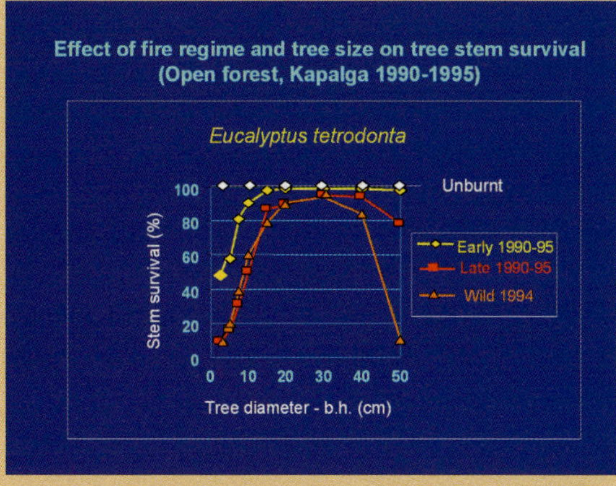

Effect of fire regime and tree size on tree stem survival (Open forest, Kapalga 1990-1995)

3   Small and very large stems are more prone to being killed, especially with more intense fires (the large stems are often hollowed, see p. 32). For many eucalypt species, stems of intermediate size (20–40 cm diameter at breast height) survive best.

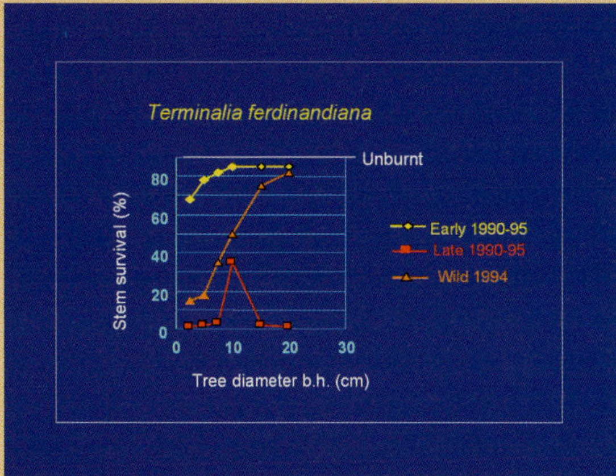

4   Stems of many small, non-eucalypt tree species, e.g. *Terminalia* spp. are sensitive to fire, especially to intense fires.

by Dick Williams

Further reading

Williams D. et al. (1998). *International Journal of Wildland Fire* **8:** 227–239.

Williams D. et al. (1999). *Australian Journal of Ecology* **24:** 50–59.

Which woody plants are most affected?

Shrubs and small trees are most affected by fire because their bark is often thinner. Many deciduous species, such as rosewood (*Terminalia volucris*), are more sensitive to fire than evergreen trees like eucalypts. Even large, old trees can be killed if they are hollowed out by termites or damaged. A hollow trunk acts as a chimney so the fire burns away from the inside.

A few species in the savanna are particularly sensitive; they are killed by fire and regenerate only, or mostly, from seed. These species require several years free from another fire for the seedlings to mature and replenish their seed store in the soil. Fires recurring at intervals shorter than their regeneration cycle will eliminate them from the community.

Fire-sensitive species include cypress pine (*Callitris intratropica*), many acacias, for example lancewood (*Acacia shirleyi*), and the acacias that are major components of pindan communities in Western Australia. They also include many species with very restricted or patchy distributions in heath communities on rugged sandstone ranges, especially in the Top End and Kimberley, and on coastal dunes.

Cypress pine is highly susceptible to intense fires

Some Acacia communities, such as lancewood, are susceptible to intense fire

Fire ladders (above) up the flammable paperbark do not often kill mature Melaleuca trees given they resprout from epicormic buds under the outer bark. However, paperbarks growing in peaty soils may be readily killed if the fires burn their roots (below)

Shrubs in sandstone ranges

There are some truly rugged sandstone ranges embedded in Australia's savannas. Their dramatic scenery is the heart of many national parks, and visitors find their extraordinary and beautiful flora bewitching. Yet few people, including some managers, appreciate the enigma posed by their vegetation. Many shrubs that are killed by fire find shelter in the rugged landscape. If burnt, they are totally dependant on surviving seed to found a new generation; this must then develop to flowering size to replenish the seed bank before another fire.

But spinifex relishes the same habitat. Spinifex is a perfect fuel because its resin-rich, needle-like leaves form well-ventilated hummocks. And it does burn, for everywhere but on the most isolated cliff ledges one can find charcoal, scorched branches or other evidence of fire at some time.

Despite this, the fire-sensitive shrubs are often seen growing profusely amongst spinifex hummocks in the sandstone.

'Bewitching flora of the sandstone ranges'

How can this be?

The secret lies in how frequently these habitats burn. Provided the shrubs can replenish their seed banks (they often take five or more years to flower and longer to shed enough seed) they can live with the risk of fire because, in this rugged landscape, bare rock creates sheltered areas. Fires do get there sometimes, from lightning or from burning leaves blowing in, but they have been infrequent.

This is changing. For many reasons and in many places, fire is now more frequent in the sandstone. The consequence is local extinction of many fire-sensitive shrubs in their former strongholds.

by Tony Start

Further reading

Russell-Smith J. et al. (1998). *Journal of Applied Ecology* **35:** 829–846.

Fire-sensitive plants survive through isolation and infrequent burning in a rugged landscape.

Grasslands of Cape York Peninsula— a fire-dependent habitat

Grasslands in Cape York are being invaded by woody plants, particularly tea-tree (*Melaleuca* spp.), in the absence of fires or under limited burning. The diagram below illustrates the effects of fires at different seasons on development of woody suckers.

Grasses compete with tea-trees through the wet season, but die off earlier in the dry season than the deeper-rooted trees. Fires cut back tea-trees, but also stimulate growth. The small amount of grass regrowth following an early dry season fire is soon grazed out or dies, while the tea-trees continue to grow.

The later in the dry season that a fire is lit, the smaller the tea-trees will be by the next wet season. Only very late dry season fires or storm-burns will keep most re-suckering tea-trees below the grass height. After four or five years with no fire or early dry season burns, the grasslands can be completely lost to tea-tree woodland.

A major implication of grassland thickening on Cape York is the ensuing loss of habitat, particularly for granivorous birds such as the golden-shouldered parrot, star finch, Gouldian finch, buff-breasted button-quail and black-faced woodswallow.

The processes involved in loss of habitat include changes in vegetation structure. This has led to more successful predation by birds such as pied butcherbirds and loss of perennial grasses such as cockatoo grass (*Alloteropsis semialata*), which seed-eating birds rely upon for food at critical periods of the year (especially the early wet season).

Vegetation thickening also results in loss of termite mounds in which the golden-shouldered parrots nest.

by Gabriel Crowley

Further reading

Crowley, G. M. and Garnett, S. T. (1998). *Pacific Conservation Biology* **4:** 132–148.

Crowley, G. M. and Garnett, S. T. (2000). *Australian Geographical Studies* **38:** 10–26.

The effects of fire in different seasons on the development of wood suckers

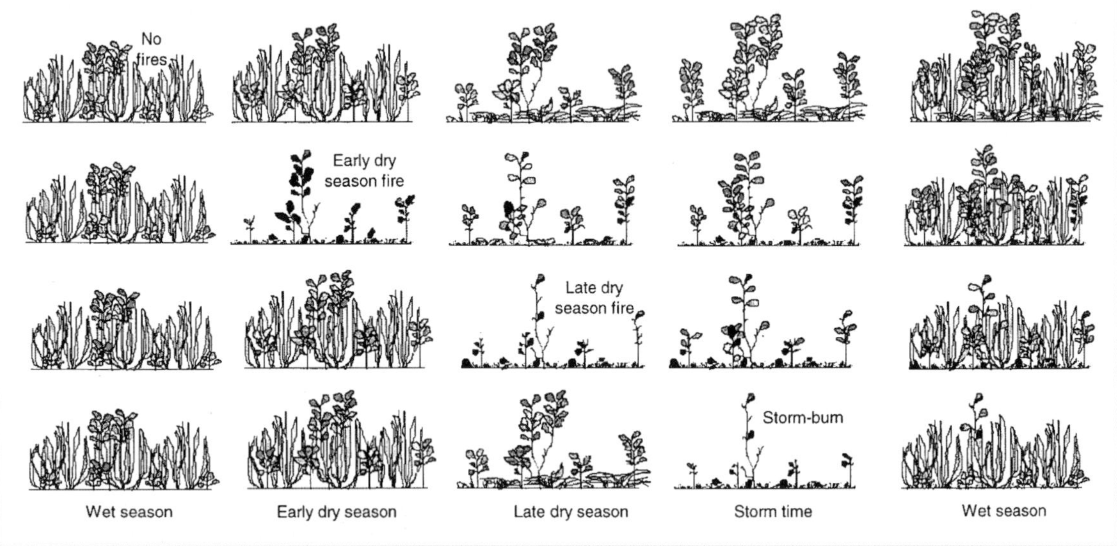

These consequences of fire are not unique to Cape York Peninsula. Except for the potential high density for tea-tree suckers, similar effects occur in many grazed and ungrazed woodlands.

How does fire affect vegetation types?

Savanna landscapes grade from open forests through open woodlands to tussock grasslands along gradients of decreasing soil moisture, soil depth or drainage capacity (see Chapter 2).

However, many sites could support any of several plant communities that differ in structure (e.g. the proportion of trees and shrubs) and composition (the mix of species).

Fire regimes often determine which plant community will inhabit a particular area. For example, occasional intense fires can generate wattle-dominated shrublands by stimulating germination of seed whereas frequent intense fires will eliminate wattles. These frequent intense fires will suit some species, such as annual *Sorghum*, which can generate sufficient fuel each year to maintain that fire regime and thus the species' dominance.

Removing fuel by heavy grazing can reduce the incidence and intensity of fires. This may lead to denser stands of woody plants because shrubs and tree saplings are not thinned out. Anecdotal and photographic evidence shows considerable thickening of heavily grazed woodland over the last five or six decades in many areas. Whilst lack of fire has played a role, other factors may be important, for example the woodlands may be recovering from dieback caused by the northern droughts of the 1930s and the 1960s (see p. 38).

How long before woody regrowth becomes fire resistant?

Without fire, some stems in higher rainfall open forest grow more than 2 m in height, and hence out of the reach of low-intensity fires, within three years. On woodland sites, however, it can take at least five fire-free years for stems to reach 2 m.

In lower rainfall regions, rosewood and other species take about 10 years to reach this height. Intense fires at intervals of less than this critical period are sufficient to reduce plants to ground level.

How does fire affect the composition of major vegetation types?

Savannas are typically very resilient to frequent burning. Annual grasses may increase in dominance with frequent burning as this can kill the top growth of woody plants—but not necessarily eliminate them.

Embedded within the savanna matrix are small but significant vegetation types occupying restricted habitats. Many of these are not so resilient, even to fires every five to 10 years. They include monsoon rainforest patches, *Acacia* scrubs, cypress pine stands, heath communities and riparian strips. Frequent fires may eliminate such communities rapidly or pare their borders gradually, year after year.

Change in a plant community's composition and structure brought about by fire will have flow-on effects to animals and other plants, particularly if a dominant species dies out, because their habitat is changed. Some characteristic responses of common community types to different fire regimes are given in Table 4.2 on page 36.

Frequent intense fires in unstocked or lightly stocked areas encourage annual Sorghum, *which leads to frequent intense fires.*

Once saplings grow above 2 m in height they tend to become resistant to fire.

Table 4.2 Characteristic responses of major savanna vegetation types to different fire regimes

Vegetation type	Environmental conditions	Woody canopy height (m)	Woody basal area (m²/ha)	Fuel types
Widespread types				
Mesic savanna, i.e. savanna types in relatively seasonally moist climates	various substrates; >900 mm rainfall pa; mostly eucalypts over annual (e.g. *Sorghum*) and perennial grasses	15–20+	7–15	grass, leaf litter
Semi-arid savanna	various substrates; <900 mm rainfall pa; plant composition as for mesic savanna	5–15	1–5	grass, leaf litter
Tussock grassland (Mitchell grass)	clay soils; low annual rainfall; mostly perennial tussock grasses (e.g. *Astrebla*)			grass
Restricted types				
Rainforest	various substrates, from ever-wet to seasonally dry; >600 mm rainfall pa; large variety of woody genera	5–25	10–60	leaf litter, humic
Acacia scrubs	various substrates; typically <1000 mm rainfall pa; sparse grass, various *Acacia* spp. e.g. *Acacia shirleyi*—lancewood *Acacia harpophylla*—brigalow *Acacia cambadgei*—gidgee	8–15	5–15	leaf litter
Cypress woodlands	various well-drained substrates; *Callitris intratropica* intermixed with eucalypts, etc.; sparse grass	10–15	5–10	fine leaf litter, grass
Heathland (including pindan shrublands in the NW Kimberley)	typically rocky or sandy substrates derived from sedimentary rocks; large variety of shrub species and hummock spinifex (*Triodia*) grasses	1–4	<1	grass, shrubs
Wetlands	seasonally inundated flood plains; paperbarks (*Melaleuca*) over perennial grasses and sedges	5–25	5–45	grass/sedge, humic

Fuels loads t/ha	Regeneration of woody plants	Response to fire regime
3–6	typically resprouters	• intense fires kill stems (typically not whole plant); some species more fire-tolerant (especially eucalypts) than others • frequent fires (e.g. annual, biennial) result in increase of annual grasses • seedlings of dominant woody species only occasionally observed—typically germinating in wet season following fire • occasional fire results in woody thickening and loss of annual grasses
2–6	typically resprouters	• as for wetter savanna • however, rate of regeneration of woody stems is slower—maybe 10 years to attain 2+ m height
2–5		• seldom burnt • frequent fires probably result in increase of annual grasses (e.g. *Iseilema* spp.—Flinders grasses)
<4	mostly resprouters	• moist, compact fuels restrict spread of fires within rainforest patches • patch margins tolerant of low–moderate intensity fires • highly sensitive to intense fires at patch margins resulting in canopy death and spread of flammable grasses • very slow spread of rainforest in absence of fire, especially on seasonally dry sites
1–2	resprouters (e.g. brigalow); obligate seeders (e.g. lancewood)	• low, compact fuel loads restrict spread of fires within scrubs • scrub margins sensitive to intense fires • canopy scorch typically results in death of obligate seeder stems, and germination of seeds in following wet season • may take 10+ years for re-establishment of obligate seeder soil seed bank on harsh sites
1–3	obligate seeders (cypress)	• much as for obligate seeder *Acacia* shrubs above, save seed bank held in cones on trees. • seedlings establish in absence of fire and very sensitive to being burnt
2–10	many obligate seeders, but mostly resprouters in extensive Cape York heaths	• rapid build-up of fuel loads in absence of fire • a regime of frequent fires (<3 years) will eliminate many obligate seeder shrubs • on some harsh sites some obligate seeders may take 5–10+ years to develop viable seed banks • vast majority of seedlings establish only in first wet season following fire
5–10	resprouters; few woody species	• rapid build-up of organic/humic fuel loads if not frequently burnt • intense occasional fires will kill mature stems by consuming roots in soils • massive seedling regeneration may occur after intense fires, given favourable flooding regime in following wet season. But, once established, saplings tolerant of frequent, low-intensity fires

More trees

Woody thickening in the VRD

Europeans began settling in the Victoria River District in 1883 and the earliest known photographs were taken in 1893. By locating and rephotographing the scenes in historic photographs we can document environmental change over almost the entire period of European settlement. This record can be supplemented with historic records and, for the past 50 years, with oral history. Enough early photographs exist to indicate the vegetation cover in the first few decades when the impact of European land management was relatively slight.

Our comparisons clearly show that in riverine environments—river and creek banks, and the plains to the base of nearby ranges—there has been a considerable increase in tree cover. Significantly, the change appears to have either begun or accelerated in the post-war period. Changes in vegetation on other landforms are far less evident although the general trend appears to be a slight increase in tree cover.

by Darrell Lewis

Part of a cycle

Episodic dieback and recovery

The evidence that the Australian landscape was substantially more open at the time of European settlement depends on the interpretation of the historical record. While some dense vegetation, such as rainforest in north-eastern Queensland has undoubtedly advanced, we conclude that broadscale conversion from grassland and open woodland to forest is overstated.

Stocks of woody vegetation can undergo considerable flux. But are these fluxes the result of normal climatic cycles, of cattle grazing, of changes in fire regimes or due to CO_2 fertilisation from increasing greenhouse gases? Dieback collapses can undoubtedly result from extreme drought events. The figure at right shows two scenarios. Scenario A portrays a system with sudden declines—in this case corresponding with the major Queensland droughts of this century—followed by increases corresponding with average or above-average rainfall. Scenario B represents the same trends but with an overall increase related to fire and other management implications of replacing Aboriginal management with rangeland pastoral management. The actual situation probably lies somewhere between these extremes.

by Rod Fensham

Further reading

Fensham, R. J. and Holman, J. E. (1998). *Rangeland Journal* **20**: 132–142.

Coolibah Station in the Northern Territory in 1960 (above) and 1997 (below); woody vegetation has thickened over 40 years.

Eucalypt dieback in north Queensland during 1990s drought

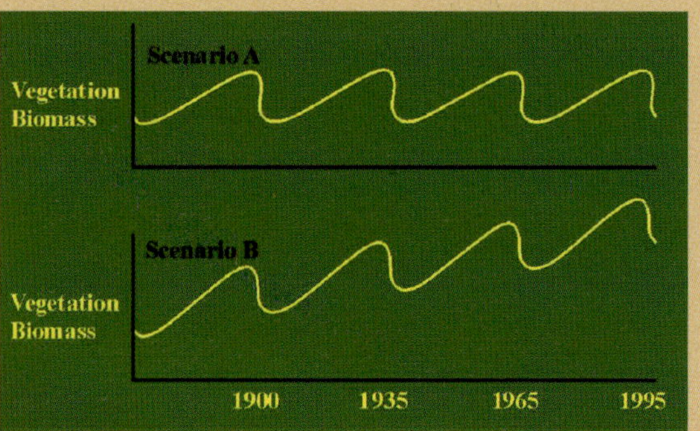

Fire and mistletoes

Mistletoes are a natural and diverse component of Australia's tropical savannas. Most species show some degree of host preference; some grow only on eucalypts, others prefer acacias.

In healthy landscapes, each species lives in harmony with its hosts while contributing nectar to honeyeaters, fruit to the mistletoe bird and palatable leaves to herbivores. They thus play an important role in the whole ecosystem.

But many mistletoe species have a peculiar problem. If the host is killed by fire, the mistletoe dies too; indeed most mistletoes will die if they are scorched, even if the host can resprout. If all the hosts in an area are burned, all the mistletoes die.

Mistletoes, then, are obligate seeders but, unlike other obligate seeders they have no seed bank to initiate a new generation. Their seeds have to be 'planted' on the branch of a suitable host by a bird. After fire has killed a population of mistletoes the only source of seed is another population, usually one growing outside the burnt area. It becomes even more drastic where the host is another obligate seeder, like many of the acacias. Then the process of recolonisation cannot begin until the hosts have grown up—usually adding several years to the process.

Frequent, extensive intense fires can eliminate mistletoes from huge areas, and this has undoubtedly happened in parts of the tropical savanna. No one knows what the effect has been for nectar and fruit-eating birds or the other plants that shared their services. However, we do know some herbivorous insects depend on mistletoes. For example, the larvae of some of our most spectacular butterflies feed on nothing else, so they too will have disappeared from huge areas.

The converse is that the presence of a diverse array of mistletoe species, including some that grow on obligate-seeder hosts, can indicate that the area is long unburnt or, at least, has had a regime of mostly mild and probably small fires. This can be a useful indicator because there are few parts of the tropical savanna where we have a long record of the fire regime. Being able to identify areas where the regime has been mild may allow us to examine many other aspects of the impact of fire over long periods of time.

by Tony Start

Further reading

M. Calder and P. Bernhardt (Eds) (1983). *The Biology of Mistletoes.* Academic Press. Sydney.

Start T. (1998). In *Fourth North Australian Fire Management Workshop. Kalumburu, North Kimberley, Western Australia, June 1997.* (Eds P. Saint and J. Russell-Smith.) Bushfires Board of WA, Kununurra, and Tropical Savannas CRC, Darwin.

Effects of fire on native animals

"We were amazed to see the effects of feral animals and weeds, and unmanaged fire in some places. We noticed that there was less variety, less biodiversity of both plants and animals in places that had been having late hot fires every year for years. We also saw country where people were trying to protect it from fire, to have lots of fruit trees and shrubs for animals, like emu especially, and people."

Dean Yibarbuk on the lack of fire management in western Arnhem Land.

How do fire regimes affect vertebrates?

Just as the structure and composition of plant communities reflect their fire history, there will be more habitat diversity in areas where there is greater variety in the frequency and intensity of fires.

The size of the different habitat patches (and thus burnt areas) is important. Many animals require a variety of habitats, for example for breeding and for foraging, and large uniform areas do not suit them. Patchy burning creates vegetation that at any one place changes from newly burnt to long unburnt.

Such habitat diversity created by numerous small fires ensures that at any one time there are habitat patches that meet the needs of each animal species. Futhermore, the habitat patches will change with time, allowing different species to colonise patches when they become suitable. Large, uniformly burnt or large unburnt areas will fail to maintain this biodiversity.

Besides influencing habitat variety, fires have other effects on the resources needed by animals. For example, fires reduce flowering and fruiting of savanna trees, particularly in the season following a late intense fire. If fires occur over extensive areas they can have a significant effect on the availability of food for nectar- and fruit-eating birds.

Fire regimes that advantage one species may disadvantage another and be inconsequential to many more.

Late intense fires reduce food for fruit-eating mammals.

Fire, fruit and emus

Eighteen thousand years ago, on a rock wall on the west of the Arnhem Land escarpment, an Aboriginal artist depicted an emu being speared by a hunter. Emu hunting has continued to present times through major changes in the landscape.

Dean Yibarbuk, chief ranger with the community-based Djelk Rangers at Maningrida, says, "Emu is very important, not just for food, but it's also an important animal in ceremony and totemic associations for some clans.

"It's an unusual bird as it is not classified in our taxonomy with other birds, but is included with large game animals like kangaroos.

"It has been a favourite food of our people for tens of thousands of years but there is now a concern that the numbers are going down.

"Perhaps the problem lies in changes in fire regimes. Too many hot fires will kill the fruit trees or stop them producing for years while they recover. Predators like dingoes or pigs could also be having an effect—or maybe it's a combination of factors."

Dean says: "People want to see the numbers of emus come back again so they can continue to have natural 'emu farms' for their own domestic use. And perhaps there may be some commercial future there as well."

by Dean Yibarbuk and Peter Cooke

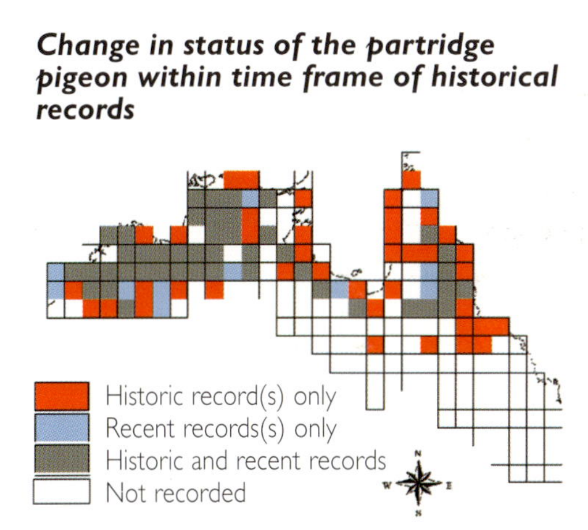

Historic and recent distribution of the Gouldian finch

Change in status of the partridge pigeon within time frame of historical records

Seed-eating birds are declining in the tropical savannas

The grassy savannas of tropical Australia are particularly rich in the variety and numbers of seed-eating birds. The 14 species of finch found in the area is without parallel elsewhere in Australia. There are also many other species including 11 pigeons, seven parrots, six cockatoos, six button-quail and three quail. Of these, five finches, three pigeons, three parrots and two button-quail are found nowhere else. Concerns that these seed-eating birds are not coping well with change have been around for a while.

On Cape York Peninsula, the golden-shouldered parrot is now endangered and very rare. In the Katherine district of the Northern Territory, the Gouldian finch almost disappeared in the late 1960s or early 1970s. In the late 1970s, finch trappers operating in the Kimberley noticed the same thing. There were rumours of other species that were not doing well, but not much evidence to support or refute these suggestions.

In 1997, the Tropical Savannas CRC combined with Environment Australia's Gouldian finch recovery project to assess this change. Historic literature and recent databases were combed for records of seed-eating birds, finding 3360 historic records prior to 1951 and over 60,000 records overall. This was far in excess of expectations. These were analysed for evidence of change in the distributions and frequency with which the species were reported.

The results were alarming. At least 11 species have declined, some of them quite drastically. Several of these were quite unexpected, such as the common bronzewing pigeon (Northern Territory and Western Australia only). Only three species have increased— galahs, crested pigeons and peaceful doves.

Why is this happening?

The answers are neither simple nor clear, but are likely to include either or both the effects of changes to burning patterns or the consequences of grazing. What is clear is that the changing pattern of human settlement that followed the arrival of Europeans in the tropical savannas has had a major effect on the seed-eating birds.

by Don Franklin

Further reading

Franklin, D. C. (1999). *Biological Conservation* **90**: 53–68.
Franklin, D. C. (2000). *Wingspan* **10**: 14–21.

Winners and losers with savanna fires

Many environments are changing rapidly, causing some species to decline, others to increase. Changes are most extreme where the current fire regime is most dissimilar to that previously operating.

The most notable example of this contrast between old and the new fire regimes is in the pastoral country of inland Queensland, the Barkly Tableland and parts of the Victoria River District, where fires are now mostly excluded over very extensive areas.

The degree of this change in fire regimes and the vast area over which it has been made mean that species or communities associated with the previous regimes have declined substantially over very large areas. When the new regimes become entrenched, as when fires occur at about the same time every year, or where fire is excluded over large areas for years, gradual change will lead to local, regional or complete extinctions

Too much for the Gouldian finch

The Gouldian finch has declined dramatically across much of its range in the northern savannas and is now regarded as endangered. This bird eats only grass seeds, with most food taken from seeds fallen to the ground. Foraging is difficult in a dense grass layer, but far easier where patchy fires early in the dry season remove the grassy bulk. Thus, Gouldian finches will move (perhaps tens of kilometres) to congregate on recently burnt areas. Without early fires the finches may not survive the dry season. But late intense fires not only remove the dense grass but also destroy most of the fallen seeds. Fires at different times (or no fires) may also affect when and how much seed is produced—potentially critical for finches to survive the wet season.

Fires dictate not only the immediate and forthcoming availability of food, but also have longer-term effects. Consistent patterning of burns repeated over many years generally lead to longer-term changes in the understorey composition and vegetation structure, to the benefit or detriment of many animal species.

Too little for the golden-shouldered parrot

The endangered golden-shouldered parrot nests in termite mounds in grasslands in Cape York Peninsula. With recent removal of intense fires, these grasslands are being invaded by dense stands of *Melaleuca*

(see p. 34). Butcherbirds can now perch close to the parrots' nests and attack and kill their fledglings. There is also less food from seed of cockatoo grass. This has lead to a great reduction in the range and abundance of the golden-shouldered parrot.

Generalised responses of birds, mammals and reptiles to different fire regimes are summarised in Table 4.3.

The grass seed-eating Gouldian finch is declining under a regime of intense late fires.

The golden-shouldered parrot is declining in Melaleuca *woodlands under a regime without intense late fires.*

Table 4.3 General response of characteristic groups of fauna in savanna landscapes to different fire regimes

Faunal group	Fire regime			
	Infrequent or no fire	**Frequent low-intensity fires**	**Frequent intense fires**	*Occasional wet season fires*
	occasional, potentially intense fires at intervals >5 years	*patchy early dry season fires at intervals <3 years*	*extensive late dry season fires at intervals <3 years*	*patchy low-intensity fires*
Birds				
fruit-eating birds	**increase** in abundance, diversity	good diversity	**decrease** in abundance, diversity	no effect
seed-eating birds	**decrease** in abundance, diversity	good diversity —best regime	**increase** in some species, **decrease** in others (e.g. finches) given poor access to seeds, and habitat changes	**increase** in some species (e.g. partridge pigeon)
insect-eating birds	**increase** in abundance, diversity	good diversity	**decrease** in abundance, diversity	no effect
Mammals				
tree-dwelling mammals	**increase**	good diversity	**decrease**	no effect
terrestrial mammals	**reduction** in grazing species, **increase** in some species (e.g. rodents, quolls)	good diversity	**decrease** in some species (e.g. rodents, quolls)	no effect
Reptiles	general **increase** in ground-dwelling lizards (e.g. skinks) given greater litter, but **decrease** in dragons	good diversity (and good for frillneck lizards)	general **decrease** in ground-dwelling lizards given loss of litter, **increase** in dragons	no effect

Frillneck lizards and fire

The frillneck lizard (*Chlamydosaurus kingii*) is a tree-dwelling species found throughout northern Australia in forests and woodlands that are regularly burnt in the dry season. To cope with the long dry season, frillnecks reduce their metabolic rate by 70% and can spend as long as three months sitting in the same tree.

Frillneck lizards respond in the short term to dry season fires. They can find more food (invertebrates) in the week after the fire, especially in the late dry season, and prefer to stay in the burnt habitat. However, intense late dry season fires can kill 30% of the lizards as they come out of their trees during intense fires to hide in old termite mounds. Early dry season fires caused no mortality.

Over longer time periods there were fewer differences among populations exposed to different fire regimes. Diet, home range and growth of individuals were similar among populations exposed to annual early fires, annual late fires and no fires. However, an area that remained unburnt for a number of years became unsuitable for the species, and net migration out of these areas resulted in very low population densities.

Using the information on population ecology, a simulation model was run on the viability of frillneck lizard population under a range of fire intensities and frequencies. Early dry season fires, even on an annual basis, pose no real threat to the persistence of frillneck populations. Late dry season fires occurring at a frequency of greater than two years in the same population greatly increase the risk of local extinction as the number killed by fire cannot be matched by the level of reproduction and immigration.

Habitats unburnt for more than 20 years also pose some risk to the species as the thick cover of plant material inhibits access to ground-dwelling invertebrates.

by Tony Griffiths

Further reading

Griffiths, A. D. and Christian, K. A. (1996). *Australian Journal of Ecology* **21:** 386–398.

Food

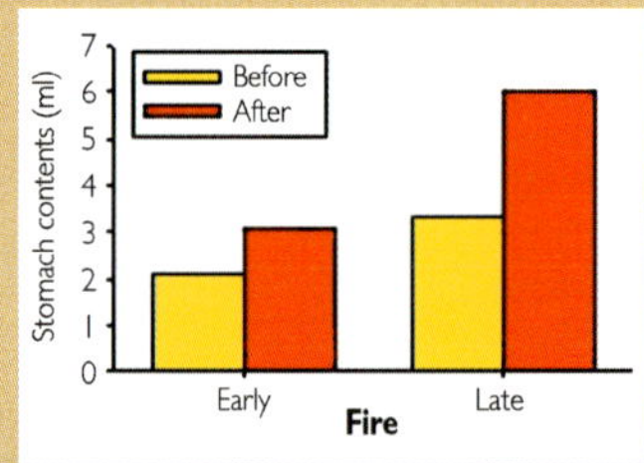

Frillneck lizards can find more food after a late intense fire, but may be cooked themselves.

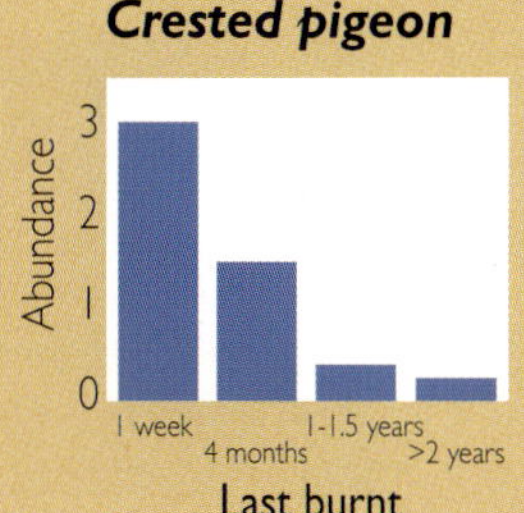

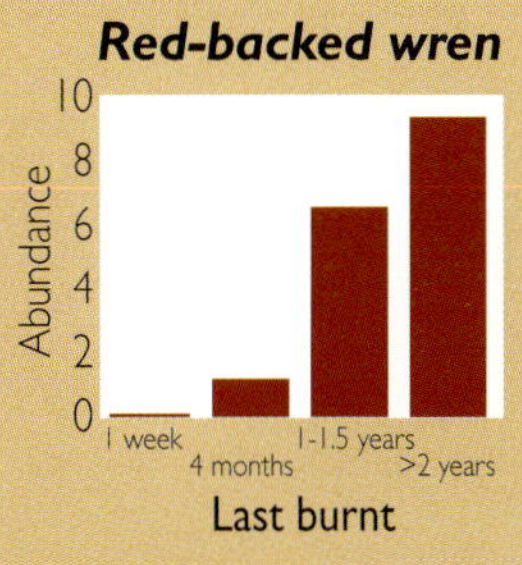

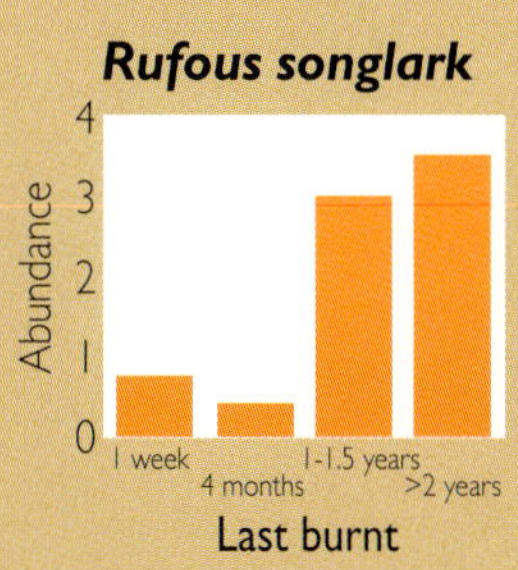

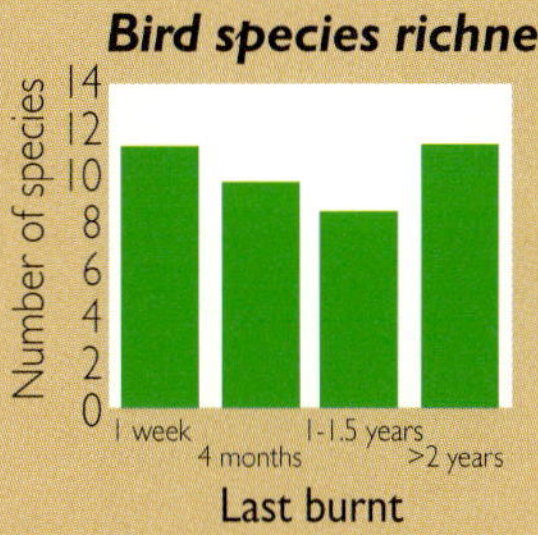

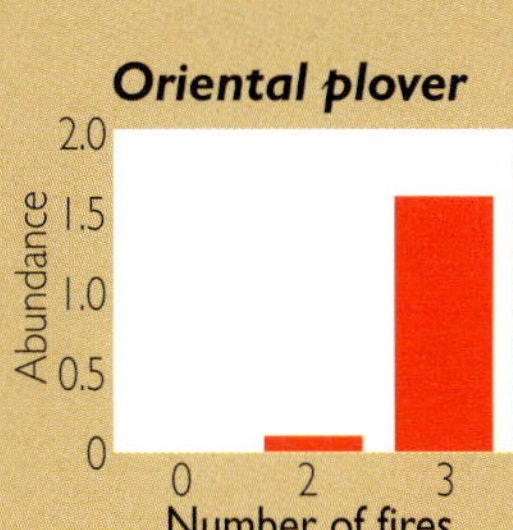

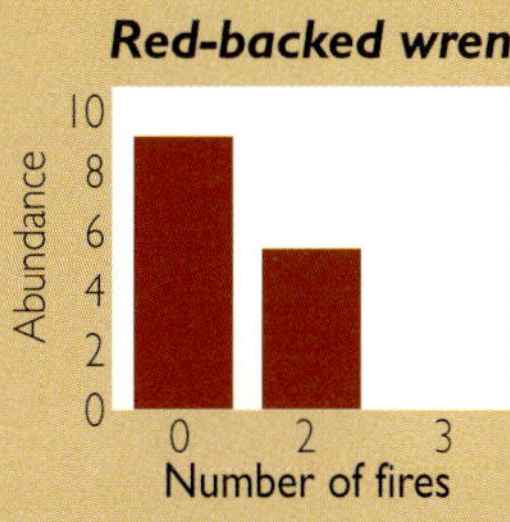

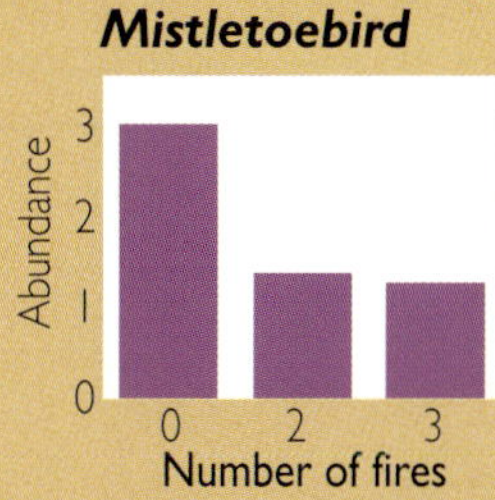

More winners and losers

Fires at a particular time of year may destroy key resources needed by one species but make available resources for another one. Thus, a fire may create areas of open ground on which doves can forage while destroying a nectar crop for honeyeaters or thickets essential to fairy-wrens. Fires may destroy the nests of some birds during the breeding season but help form the tree hollows needed for breeding by parrots or shelter by owls and bats (but fire can destroy tree hollows). Whatever an individual species may need, habitat variety created by a range of fire regimes is necessary if an area is to cater for the needs of them all.

For any consistent fire regime, there will be species which are benefited and species which are disadvantaged. The examples on the left are from a five-year experimental fire study at Kidman Springs in the Victoria River District.

Here, diamond doves and crested pigeons occur mainly at sites which have been burnt recently, whereas red-backed fairy-wrens and rufous songlarks prefer the dense undergrowth occurring in sites long unburnt.

Bird species richness is greatest at both the most recently burnt sites and the sites longest unburnt, probably because one group of species favours recently burnt areas and another group occurs only in, or are most abundant in, long unburnt areas.

Fire regimes are a complex of factors, and it may not be merely time since the last fire which is critical for particular animal species. Other important attributes include fire frequency, intensity, extent and timing. For example, the oriental plover and zebra finch prefer sites burnt often whereas the red-backed fairy-wren, mistletoebird and the blind snake (*Ramphotyphlops guentheri*) are all disadvantaged by frequent fire.

by John Woinarski

Further reading

Woinarski, J. C. Z. (1990). *Australian Journal of Ecology* **15:** 1–22.

Woinarski, J. C. Z. et al. (1999). *Rangeland Journal* **21:** 24–38.

How does fire affect insects and other invertebrates?

Fire often has little long-term effect on the overall number of insect and other arthropod species or on their abundance, but it can change the mix of species in an area. The question 'How long ago was the last fire?' is more important than 'In what season did it occur?' For example, grasshoppers (like many other insect groups) are influenced more by rainfall than by fire. However, season of burn is important to many ground-active insects, such as ants, which are affected more by intense late season fires.

Management for conservation of invertebrates should focus on maintaining patches of savanna unburnt for three to five years rather than worrying too much about season of burn. Once again variety in fire regimes is important.

by Alan Andersen

Fire and ants

The main environmental aspects influencing ant communities include litter cover and vegetation structure. Changes in the levels of these may allow different types of ants to dominate.

The effects of fire on ant community structure are illustrated in the graph showing the responses of

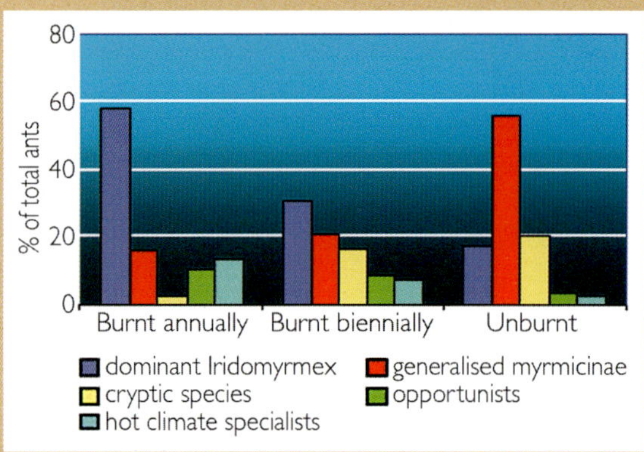

some ant groups to different experimental fire regimes in Kakadu.

The open conditions produced by annual fires favour the dominant meat ants (*Iridomyrmex*). 'Opportunists' (*Rhytidoponera* and *Tetramorium*) and hot climate specialists (*Melophorus, Monomorium* and *Meranoplus*) were also abundant. Generalised myrmicinae, although favoured by open, insolated conditions are limited in this case by competition with *Iridomyrmex*.

Litter that accumulates when fire is excluded provides vital habitat for cryptic species (i.e. those that conceal themselves) that nest and forage predominantly within soil and litter (many genera, including *Pyramica* and *Strumigenys*). Cryptic species do not interact greatly with other ants. These same environmental conditions are less suitable for the dominant *Iridomyrmex*. Freed from competition with the *Iridomyrmex*, the generalized myrmicines (*Monomorium* and *Pheidole*) increased substantially in abundance.

Ant communities from plots burned every two years were on average intermediate between the two extremes of annually burned and unburned plots.

Further reading

Andersen, A. N. (1991). *Biotropica* **23:** 575–585.

CSIRO

Grasshoppers

The spectacular Leichhardts grasshopper (*Petasida ephippigera*) is endemic to sandstone regions of northern Australia and is totally dependent on *Pityrodia* spp. as host food plants. Burning its habitat in the dry season may affect local populations of *Petasida* as the wingless grasshopper nymphs cannot escape extensive fires.

by Lyn Lowe

Lyn Lowe

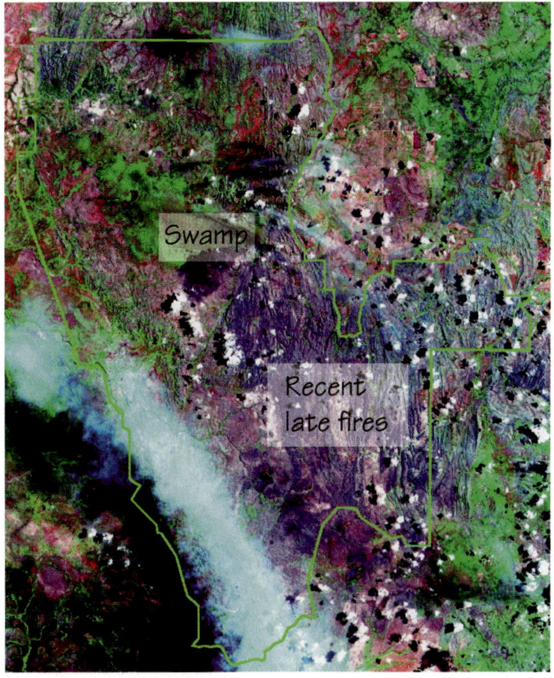

Landsat image of Litchfield National Park in September 1997. The light-blue band in the south-west is a smoke plume blown across the park by the typical dry season south-easterlies. The rolling hills in the mid-east have been recently burnt (dark blue). As the ash blows away over time, bare earth becomes highly reflective (bright red, fading to pink). The spring-fed Tabletop Swamp remains wet and green.

Figure 4.2 Fire map of Litchfield National Park from Landsat imagery

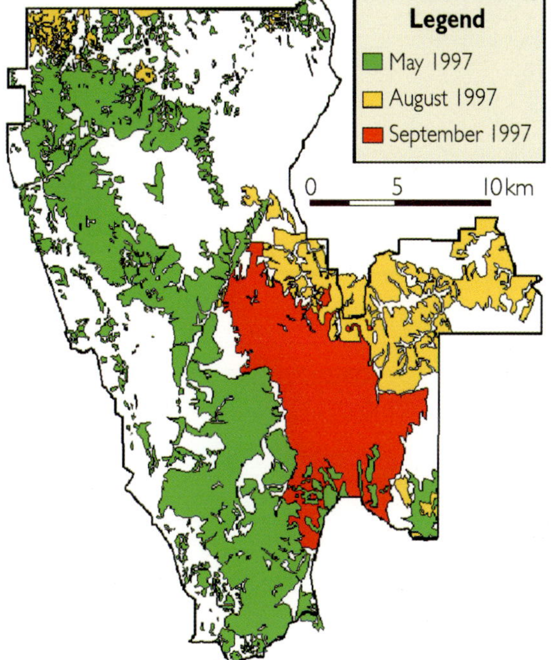

Fire map of Litchfield National Park, NT, from Landsat imagery. Early fires (green) are more patchy whereas burns become more complete by the late dry season (red).

What's an ideal fire regime for animals?

A fire regime that will optimise animal diversity requires variety and variability, preferably at a fine scale, so that individuals can pick and choose areas that meet their various requirements. Equally important, in a fine-scale patchy mosaic of habitats there will always be populations that can colonise areas as the vegetation changes through time between fires. In the huge, frequently burnt areas that characterise much of the savanna landscape today we need to retain and sustain some long unburnt areas and allow others to develop. In the rarely burnt prime pastoral lands of central Queensland, the Barkly Tablelands and Victoria River District we need to reintroduce some burning.

What is an optimal area to burn?

The examples in this chapter demonstrate that a large area of monotonous habitat will cater for fewer animal species and lower abundance of many of those that are there, be they mammals, birds or reptiles, than will the same area if it contains a variety of habitats generated by fire. However, because some groups of animals are more mobile than others (for example birds versus lizards) it is not possible to generalise on the optimum size for habitat patches (and thus burn areas).

Nevertheless, a mosaic of smaller patches will provide better habitat for many more species than one of larger patches.

What's in the air?

When savannas burn, more than 90% of grassy fuels is given off as smoke. This smoke consists of water vapour, a range of other gases and very fine particulates. Most of the gases are various oxidised forms of nitrogen or carbon, with carbon dioxide dominating. However, a similar amount of carbon dioxide will be absorbed into the next season's growth of vegetation and it is assumed that the net production is zero.

Methane emissions, although a small percentage, are important as a greenhouse gas. Little is known about methane production from decomposing litter (as opposed to combustion of fuel/litter) in northern Australia. Fine particulate carbon—less than 10 millionths of a metre in size—is important for its potential effects on human health. Nitrogen emissions are dominated by elemental nitrogen, which is produced by combustion of amino acids in plant tissues. Ammonia and various nitrogen oxides are also major nitrogen-based gasses released from fires.

The fuel that is not given off as smoke is nonetheless charred to varying degrees as ash. Of this ash, about three-quarters stays on the ground and the remainder is carried up with the smoke plume to be deposited many kilometres away.

The ash that remains on the ground is enriched in all the major plant nutrients except nitrogen, and these are in forms more readily available to plants than in unburnt fuel. After early dry season fires the ash is blown by winds and tends to be captured by obstructions in the landscape such as grass tussocks and fallen logs. After late dry season fires the ash can be moved by run-off waters and washed to lower parts of the landscape.

An important component of the ash is elemental carbon—also called black carbon. This carbon is believed to be extremely stable in the environment and therefore represents an important store of carbon.

The gaseous losses of nitrogen in a typical Top End fire represent about 10 times the amount deposited in rainfall each year. For a frequently burnt ecosystem to remain in balance this lost nitrogen must be replaced by biological fixation. However, legumes are a relatively minor component of Australian savannas. Fixation associated with grasses or termites are other possibilities, but the amounts of nitrogen fixed in these ecosystems are unknown. The nitrogen content of rainforest soils, where fires are very infrequent, tends to be higher than those of adjacent frequently burnt savannas.

by Garry Cook

Further reading

Cook, G. D. (1994). The fate of nutrients during fires in a tropical savanna. *Australian Journal of Ecology* **19:** 359–365.

Cook, G. D., Hurst, D. F. and Griffith, D. W. T. (1995). *CALMscience Supplement* **4:** 123–128.

Hurst, D. F., Griffith, D. W. T. and Cook, G. D. (1994). *Journal of Geophysical Research* **99(D8):** 16441–16456.

Gas emissions

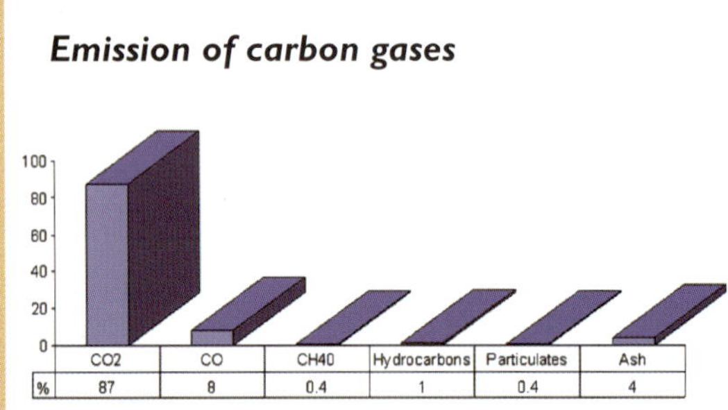

%	CO2	CO	CH40	Hydrocarbons	Particulates	Ash
%	87	8	0.4	1	0.4	4

Carbon dioxide is the main gas produced by burning but this will be reincorporated into new vegetation growth next year.

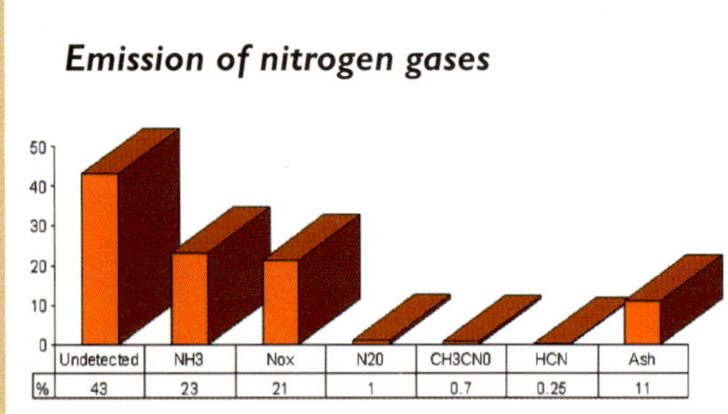

%	Undetected	NH3	Nox	N20	CH3CN0	HCN	Ash
%	43	23	21	1	0.7	0.25	11

Ammonia in smoke makes the eyes sting.

Burning resin-rich spinifex generates plenty of smoke.

What's in the water?

Burning late in the dry season (September) leads to increased soil erosion and at least a doubling of the amount of sediment carried by streams. In contrast, fires lit early in the dry season (May–June) have catchment soil losses similar to catchments left unburnt. These conclusions are based on work at Kapalga in Kakadu National Park where there was no human activity in the catchment. Soil losses caused indirectly by burning could be much more if grazing, vehicles or other activities disturb the soil's surface.

The reason for these differences lies in the way fire affects the vegetation and ground cover of a catchment. Vegetation and ground cover help to protect the soil from erosion by raindrops and water flowing over it. Fires lit late in the dry season kill young and mature trees reducing the amount of canopy cover and the stability of stream banks. Late dry season fires also burn almost all leaf litter and dead grass to expose bare ground.

Streams in the Top End usually flow only during the wet season, usually between January and June, although spring-fed streams may flow all year round. In catchments burnt late in the dry season, heavy rainfall during the 'build-up' months can result in water flowing quickly over the land to streams which then flow for 1–2 days. These first flows of water—the 'first flush'—carry large amounts of sediment, ash, nitrogen and phosphorus, and reduce stream water quality. When the sediment mixes with river, billabong or dam waters, it can even kill fish by depleting oxygen. 'First flushes' occur less frequently in catchments that have been burnt early in the dry season, or left unburnt, because of better vegetation cover.

Although water quality of 'first flushes' after a late fire is poor, stream water quality improves (but sediment load is still high) as the wet season progresses and catchment vegetation grows.

At Kapalga, the amount of nitrogen and phosphorus carried by streams during the 'wet' is similar for catchments burnt early and late in the dry season, reflecting the very low fertility of these soils. Nutrient levels would be expected to be higher from more fertile soils. Over many years the amount of nutrients carried by 'first flushes' from catchments burnt late in the dry season would reduce the fertility of the catchment soils.

by Simon Townsend

Further reading

Townsend, S. A. and Douglas, M. M. (2000). *Journal of Hydrology* **229:** 118–137.

Stream water quality of 'first flushes' and wet season storm water for catchments burnt early or burnt late in the dry season

	Stream water quality of first flushes from a catchment burnt late (September) in the dry season	Wet season water quality of storm runoff draining in a catchment burnt late (September) in the dry season	Wet season water quality of storm runoff draining a catchment burnt early (June) in the dry season
Sediment (mg/L)	390	58	9
Phosphorus (µg/L)	110	13	9
Nitrogen (µg/L)	2400	330	360

First flushes from early storm rains carry high levels of sediment off land burned late in the dry season.

5. Using fire to manage savanna

by Rodd Dyer, Jeremy Russell-Smith, Tony Grice, Tim McGuffog, Peter Cooke and Dean Yibarbuk

The previous chapter outlined some of the major impacts on the landscape of fire. This chapter describes how fire can be used to better manage these landscapes.

To ensure that fire management is effective and sustainable good planning must be based on clear objectives and ecological principles. This chapter provides some general, ecologically based guidelines for those interested in using fire to achieve particular land management goals in the tropical savannas.

"It's not just people who light fires," says Aboriginal artist Billy Yalawanga from central Arnhem Land. Yalawanga's painting shows Karrakanj, the brown falcon, picking up a smouldering stick from a burned patch of ground and dropping it into dry grass on the other side of a road. "It doesn't matter if a creek or road pulls that fire up, because Karrakanj is hungry for more insects from the fire. Karrakanj lights a new fire so he can get more food." Across northern Australia, many Aboriginal people, and some non-Aboriginal people, say they have seen this happen.

Traditional burning

Fire has been used by Aboriginal people for managing habitats and food resources across northern Australian over millennia. It needs to be appreciated, however, that such directed burning forms only part of the broader usage of fire in traditional Aboriginal society—fire was used for a variety of purposes ranging from signalling, cooking and hunting, through to meeting a range of social and religious obligations. For example, the traditional widespread practice of burning the country from early in the dry season can be understood in a number of contexts: clearing away rank undergrowth after the wet season in order to make walking easier; for encouraging regrowth of grasses ('green pick') for kangaroos and other game; for protecting certain areas and, just as fundamentally, for 'cleaning' the country in order to meet spiritual obligations.

When describing traditional burning it must be appreciated that such practices varied in different areas of northern Australia, probably most pronounced between higher rainfall coastal and sub-coastal areas and the more arid interior. While traditional burning practices are relatively well documented (e.g. in parts of Arnhem Land in the Northern Territory) as these practices continue in some locations, for most of northern Australia and especially inland areas, much of that traditional knowledge has been either lost or remains undocumented. The following example of traditional burning from the Cadell River region of north-central Arnhem Land is related by Dean Yibarbuk.

Traditional use of fire in central Arnhem Land

The secret of fire in our traditional knowledge is that it is a thing that brings the land alive again. So we do not necessarily see fire as bad and destructive—it can be a good thing and bring the country back to life. But it is not a thing to play with unless people understand the nature of fire.

A child's first experiences of fire are being with adults and seeing fires in the landscape (*manwurrk*) and also the fire that is the centre of each family living area (*kunrak*). This learning by experience and experimentation is always going on with either the parents or other close family keeping a watchful but relaxed eye on what is happening. But as they grow, young people learn that fire is more than just something for cooking or hunting—that it has a deeper meaning in our culture. As they attend ceremonies with their parents they see and learn to respect the sacred fires that are central physical parts of the most sacred of ceremonies.

In the upper Cadell River area, there is a tradition of young men's initiation involving circumcision at about the time of puberty. Fire enters into this ritual also when the new young man is 'burned' with heat, smoke and steam. At the same time he is told how to behave and is warned against things that are forbidden.

Young boys and girls participate in seasonal burning on flood plains to help the hunt for goannas, rats, snakes, bandicoots, wallabies and freshwater turtles. When elders see it is the right time for burning on the plains they start to talk together. The burning is not carried out by just one clan. Neighbouring groups get together and cooperate. These relationships are also reflected in kinship between members of the groups. The burning is not just of economic significance. It also has a spiritual purpose in making the country clear of spiritual pollution which follows a death amongst the landowners. Criteria which determine when the burn will happen include:

- judging when the grass will burn easily, but still retains some moisture so that it doesn't get too hot. This is usually at the time we call *yekke*, or cold weather time around June or July;

- wind direction is watched to make sure the fire will go wherever they want it to go. If it is flood plain burning they will keep it on the plain and drive it towards wet areas so that the remaining areas can be burned later;

- on the flood plains there are often two periods of burning. The first is an early burn at the edges while the plains are still moist and the grass is green, and then a later burn when those grasses are dried. Because of the early burn the fire is kept out of the forest areas by this traditional firebreak.

Both men and women work together for the burning but with men moving faster and over greater distances lighting fires and women carefully coming along gathering the resources revealed on the burned areas. But there is one kind of burning which is men's business alone—and it is dangerous work. This is the fire drive mainly for larger kangaroos, wallabies and emus.

There are special places where these hunts traditionally occur. In old times they happened every year. When the most senior landowner for the area where the fire drive is to be held sees that the time is getting close, he will talk with a close kinsman who, under customary law, has the responsibility for managing that area and the fire drive itself.

They discuss how the manager (or *djungkay*) will organise the drive—where and when it will be held and who will be invited. Messengers move out to where other family groups are living to carry the invitation.

The shoulder patches of the Manwurrk Rangers from the Mann River in Arnhem Land tell a story about the importance of fire in maintaining healthy populations of desirable food species such as emu. The story begins with a reference to the origin of emu, or Wurrbbarn as a species. The emu was once a greedy old woman, shown in the painting with her digging stick. Because of her greed she became the emu whose long neck represents the digging stick. The painting also shows flowers and fruits which are the staple food of Wurrbbarn as well as the two sticks which make up a fire drill or kundjakol in Kunwinjku language. Fire is needed to reduce fuels so that there are not periodic huge wildfires, but burning needs to be done in a way that it does not prevent the flowering and fruiting of plants which are emu food.

As they go, they burn unwanted grass to send up a signal of their approach. The messengers and the invited groups then head for the location for the fire drive, again sending up smoke that marks their travel.

When all the groups have assembled and have camped a night or two and made their plans, a start will be made very early on the appointed morning. The group splits into perhaps four groups each with a couple of men. Two go in one direction and the others in the opposite, circling around until they get into prearranged positions in a sort of horseshoe shape. According to plan, one fire is lit and others, seeing the smoke, then start walking and lighting the grass as they go. When the semi-circle of flame is lit, the kangaroos are driven to where another large party of maybe 10 or 20 men are waiting with shovel-nose spears.

After the hunt the hunters come back together. They use smoke from ironwood leaves to ritually purify the game so that it may be eaten by women and children. This is necessary because the fire drive is itself regarded as a sacred and very serious act, often first enacted by the major creative beings for that area. For a young man, the spearing of his first kangaroo at an event like this is very important.

Today fire is not being well looked after. Some people, especially younger children who don't know better or who don't care, sometimes just chuck matches anywhere without thinking of the law and culture of respect that we have for fire. This is especially true for people going for weekends away from big settlements. Fire continues to be managed well around the outstations where people live all the time.

The other big problem is large areas of country where no one is living permanently now. Grasses and fuels are build up, sometimes over a number of years, until one day someone's little hunting fire or a cigarette chucked out of a Toyota gets going and hundreds or thousands of square kilometres get burnt out in very hot fires. To go forward we need to encourage our children in the ways of the past. Fire must be managed and people must be on their country to manage that fire.

Reference

Yibarbuk, D. (1998). Notes on traditional use of fire on the upper Cadell River, In *Burning questions: emerging environmental issues for indigenous peoples in northern Australia,* by Marcia Langton. Centre for Indigenous Natural and Cultural Resource Management, Northern Territory University, Darwin. pp. 1–16.

Beginnings of a partnership with business. Adrian Ashley and Danny Brumel use a quadbike donated by the Rio Tinto Aboriginal Foundation to carry out early burning near Bulman in central Arnhem Land. Rio Tinto and other mining companies across northern Australia are actively engaging with landowners to discuss collaborative strategies for managing fire in the savannas.

T Mahney

Traditional burning and natural resource management

A key question concerning the traditional use of fire is the long-term impact this has on plants and animals. In much of northern Australia this question can no longer be addressed given the cessation of traditional modes of management.

However, in 1997 an assessment of the ecological effects of an essentially unbroken tradition of burning was undertaken on a clan estate (i.e. country belonging to one family group) on the upper Cadell River in central Arnhem Land. The project was undertaken as a collaborative ranger training exercise between the Bawinanga Association and the Parks and Wildlife Commission of the Northern Territory.

Traditional burning practice on this clan estate includes lighting mostly small fires throughout the year, and cooperating with neighbouring clans in planning and implementing burning regimes. The assessment was undertaken over 10 days in September after prescribed burning had mostly been completed.

Ecological assessments included: mapping of the resource base of the estate from both traditional and ecological perspectives; aerial survey of the extent of burning, distribution of fire-sensitive cypress pine, rock habitats, and a range of kangaroo and other animal resources; animal inventory; detailed ecological assessment of the status of fire-sensitive vegetation; and on-ground assessment of the intensities of observed fires. As well, ethnographic information concerning traditional fire management practice was documented in interviews with senior custodians.

Major observations included:

- A large proportion of the estate had been burned during the year of study.

- Resultant fires had been almost invariably patchy and of low intensity given that fuel loads comprised mostly leaf litter and little grass.

- Burned sites attracted important animal food resources such as large macropods.

- Important plant foods remained abundant.

- Fire-sensitive communities were well represented (e.g. cypress pine woodlands, sandstone heath, riparian rainforest communities).

- exotic plants were not recorded.

- diversity of vertebrates and fire-sensitive plants was high, including rare or range-restricted species.

The above assessment provides a rare insight into the long-term effects of traditional burning. Certainly, the particular clan estate was shown to be in better shape than many other comparable areas in northern Australia where, today, fire regimes are dominated by extensive, typically intense fires burning mostly grassy fuels. The assessment shows the value of intensive fire management over a relatively small area (about 90 sq. km), but there are obvious logistical difficulties when applying such fine-scale management over larger areas.

Further, despite evident differences in purpose between traditional indigenous and contemporary fire management practice, this assessment shows that outcomes from well and consistently executed traditional practice can be consistent with many widely shared land management and conservation goals.

by Jeremy Russell-Smith

Reference

Yibarbuk, D. et al. (2001). *Journal of Biogeography* **28:** 325–344.

Further reading

Crawford, I. M. (1982). *Records of the Western Australian Museum*, Supplement No. 15.

Jones, R. (1980). In *Ecology in Savanna Environments.* (Ed. D. Harris.) Academic Press, London. pp. 107–146.

Russell-Smith, J. et al. (1997). *Human Ecology* **25:** 159–195.

Sandstone country in the upper Cadell River area, central Arnhem Land. Part of the clan estate where this assessment of traditional burning was undertaken.

Using fire to manage pastures

Fire has a number of uses in the management of pastoral lands, some that are also relevant to conservation and traditional Aboriginal lands, albeit in the absence of significant grazing. As pastoralism is the dominant land use throughout northern Australian savannas, the use and impact of fire on pastoral lands is particularly important. There are trade-offs between using grass for grazing or as fuel for fire. Grazing also affects the amount and distribution of fuel throughout the landscape, potentially reducing the incidence, intensity and extent of fires.

Pastoralists and graziers want their cattle and grasslands to be in good condition. They can use fire to manipulate the composition and 'health' of pasture species and forage quality. They can also use it to even-out grazing across a paddock.

There are a number of potential risks in using fire. Failure of significant follow-up rainfall combined with heavy and sustained grazing pressure may result in serious damage to pastures. There is also evidence that wet season burning of steep slopes promotes erosion and loss of topsoil. Fire therefore must be used with care; there are situations when the use of fire is not appropriate.

How pasture responds depends on its condition (proportion of perennial and annual species), when and how often it is burned, on follow up rainfall and on grazing management after the fire.

Using fire to improve pasture vigour and quality

When grass growth is not grazed the old leaves become rank and accumulate; they can also become blackened by moulds following heavy dew or rainfall.

Burning old grass keeps pastures in good condition and makes fresh growth available to grazing animals.

Nitrogen is tied up in these old leaves and roots, and the plants become moribund. New grass seedlings cannot establish through the bulk of dead leaf.

The whole pasture becomes unattractive to both domestic and native herbivores.

Burning off old leaf can stimulate new growth and make it available to herbivores, improving overall diet quality. Protein levels in the fresh green leaf of the regrowth are higher, animals are attracted to the green feed and production is boosted.

When to burn?

Burning perennial grasses during the dry season when they have seeded and are dormant has no detrimental effect. But burning them early in the wet season when they are actively growing from stored reserves may weaken them—especially if followed by heavy grazing.

Unless the fire is to promote specific changes in species composition, burning is best carried out when atmospheric conditions and fuel moisture levels are less extreme—at the end of the dry season, either just before or after the first storm rains.

How often to burn?

The optimum frequency of burning will be determined by the accumulation rate of fuel. This will depend therefore on pasture type, seasonal rainfall and grazing pressure. Tall-grass pastures in high rainfall areas such as the Katherine region should be burned every 3–5 years to maintain pasture vigour. Productive pastures in more arid regions, such as Mitchell grass, may need burning only once in 5–15 years.

Using fire to manage species composition

Fire, with grazing management, can be used to manipulate the composition of pastures by linking the timing of burning to weakness in the life cycle of a particular species.

- Frequent burning during the dry season in high-rainfall regions often results in grassland dominated by tall annual *Sorghum*—which in turn encourages hot fires. But opportunistically burning annual *Sorghum* in the wet season before it drops mature seed can reduce the stand of *Sorghum* and allow more desirable perennial grasses to increase (provided their seeds are present).

Burning to control feathertop wiregrass

Why is feathertop a problem?

Feathertop wiregrass is a problem plant throughout the Mitchell grasslands for both wool and beef producers. The dart-like seeds penetrate the staples of wool, and prevent it from being combed out. This contamination costs the wool industry up to $10 million every year. Forage value of feathertop is also low, degrading the value of cattle pastures. Feathertop can dominate otherwise productive Mitchell grass pastures.

The effects of fire

In Queensland feathertop wiregrass can be controlled with burning followed by 2–3 months of dry weather. Just as a drought will kill this relatively shallow-rooted native grass, the combined stresses of burning and a lack of moisture leads to a 50–75% kill and a reduction in the size of plants.

The optimum time for controlled burning is July or August as this maximises the chances of dry conditions following the burn. A clean burn is needed from a moderately hot fire, followed by a 2–3 month dry spell (to kill the feathertop), with light grazing over an average summer for pasture recovery. The conditions at the time of burning also need to be dry—if there is moisture within the top 30 cm of soil, the feathertop will be able to respond following the burn. Kill rates with moist soil (or rain within six weeks) can be as low as 10–30% .

The use of improved long-range weather forecasting may provide the key to strategic burning. Controlled burning in late July or early August when the Southern Oscillation Index is in a rising phase would enhance the probability of receiving sufficient summer rains for pasture production, but provide the additional post-burn dry weather stress to keep feathertop under control.

by David Phelps

Further reading

Phelps, D. G. and Bates, K. N. (1996). Proceedings 9th Australian Rangeland Conference, Port Augusta, Australia.

Phelps, D. G. (1999). Proceedings VI International Rangeland Congress, Townsville, Australia. pp. 253–24.

Feathertop (Aristida latifolia)*, showing the distinct curving of the seed heads and sparse leaf production*

Dry season burning followed by dry conditions can reduce feathertop wiregrasss problems in Mitchell grass pastures.

- In tropical tall-grass pasture, kangaroo grass (*Themeda triandra*) is favoured by less intense, early dry season fires whereas black speargrass (*Heteropogon contortus*) is promoted by high-intensity, late dry season fires.

- Burning stable perennial ribbon grass (*Chrysopogon fallax*) pastures on cracking clays in the Victoria River District too frequently (every one or two years) will increase the proportion of annuals such as Flinders grass.

- Burning semi-arid short-grass pastures on red soils dominated by fragile annuals and biennials such as limestone grass (*Enneapogon* spp.), fairy grass (*Sporobolus australasicus)* and native couch (*Brachyachne convergens*) can destroy any seed reserves on the soil surface, resulting in even sparser stands in the next year.

- In the black speargrass country of eastern Queensland, a combination of annual spring burning and deferred grazing will promote speargrass over poorer quality, unpalatable grasses.

- Burning Mitchell grass (*Astrebla* spp.) tussock grasslands in Queensland in the dry season can reduce the proportion of feathertop (*Aristida latifolia*) (see p. 55).

Using fire to reduce patchy grazing

Fire can be used to even up grazing pressure within large paddocks. It can complement other ways such as extra fencing or water points and placing supplements.

Why is uneven grazing a problem?

Cattle tend to select palatable pasture species and prefer to graze certain parts of the landscape; they keep returning to the shorter, younger herbage in previously grazed patches. For example, in the VRD cattle prefer the short annual limestone grass pastures on red soils to the perennial pastures on heavier clay soils during the wet season. Heavy grazing in these preferred patches rapidly reduces growth of the better species and can degrade them; less desirable species invade and bare, scalded areas develop. Meanwhile, the rank herbage on ungrazed areas is wasted.

How does burning even–out grazing?

Since cattle will preferentially graze green feed on recently burnt areas and show little selection between species early in the growing season, burning old rank pasture encourages cattle into those areas. This can improve pasture condition throughout the entire paddock and may improve livestock performance.

How much of the paddock to burn?

The area to be burned is determined largely by the proportion of ungrazed pasture in a paddock, but it should be large enough to prevent localised overgrazing and pasture degradation. Conversely, burning too large an area will be less effective at encouraging even grazing.

Ungrazed areas can be burned on an ad hoc basis or in a regular rotational burning program, burning 20–50% of the paddock each year.

How often and when to burn?

Guidelines for burning to even out grazing distribution are much the same as for burning to improve pasture vigour and quality.

Stands of coarse annual Sorghum can be weakened by burning in the wet season before the grass has set seed.

With frequent burning, Flinders grass increases in ribbon grass communities on cracking clay soils.

Fresh growth after a fire will encourage stock onto ungrazed areas.

Burning on black soil on Flora Valley Station

Heytsbury Beef's Flora Valley Station, 120 km east of Halls Creek in the Kimberley region of WA, runs 10,000 breeders on an area of 6500 sq. km. About half the property is dominated by spinifex, and half is Mitchell grass on black soil downs. Rainfall declines in a gradient across the property, averaging 560 mm around the homestead to 355 mm in the southern part.

Wildfires, driven by strong south-easterly winds from July onward, are a major threat to the more productive downs country every year.

Burning on black soil country

Wayne Bean, manager, and Ian Hoare, head stockman, burn black soil country every 2–3 years to reduce the risk of wildfire, remove old rank growth and to promote healthier pastures.

They light many fires opportunistically during mustering, trying to burn patches that were not burnt in the previous year. This provides a natural firebreak as areas burnt in one year will not burn again in the next.

They prefer a total burn—best achieved when in warm and windy weather—lighting up early in the day and using previously burnt areas to prevent fires from travelling too far. After good rainfall (from November onward), the Mitchell grass in the burnt areas recovers to produce 'obviously better pick and feed'.

Fire also reduces localised heavy grazing, particularly around the major watering points, as stock move out across the whole paddock to seek fresh growth on burnt areas.

Burning black soil should not be undertaken lightly, and other good management practices have to be in place. Stock numbers must match the area that is burnt to prevent overgrazing of fresh regrowth and damage to perennial tussocks.

Burning the spinifex country

The red hilly spinifex country on Flora Valley is also burnt regularly to prevent wildfires, promote new fresh growth as a food reserve and to keep the country open for easier mustering by reducing prolific growth of woody species.

Before fire was used on Flora Valley the station was one of the poorest performing properties of Heytsbury Beef in the Top End. Since Wayne has incorporated burning with his grazing management, Flora Valley experienced the highest weight gains of all properties—the same property, the same cattle, different fire regime.

by Andrea Johnson

Burning Mitchell grass on black soil country every few years freshens up the pasture and encourages grazing away from water points

Burning curly spinifex on the hills provides useful wet season feed

Why burn at the end of the wet season?

For green pick

On heavier soils, grass continues to grow long after the rains have stopped. Some managers burn late in the wet season to force green pick from stored soil moisture. However, if this green pick is grazed too heavily, sensitive species will be weakened.

Burning to promote green pick at this time is generally not recommended, but if it is done it should be on an area large enough to prevent stock concentrating and overgrazing.

As hazard reduction

A partially burnt area creates a firebreak against runaway, late dry season fires. Some managers burn around a paddock in which they want to reduce woody regrowth; the inside area is spelled, then burned with a hot fire at the end of dry season.

Increasing biodiversity

Late wet season fires create a mosaic of burnt and unburnt grass that provides a variety of habitats and feed supplies for native birds and animals. When the aim is to improve biodiversity, it is best to remove cattle from the paddock for a time.

Establishing and maintaining stylo pastures

Seca and verano stylo (*Stylosanthes* spp.) are hardy legumes sown extensively in open eucalypt woodlands throughout northern Australia to boost the quality of native pastures.

The simplest method of establishing stylo on sandy-surfaced soils is to burn at the end of the dry season and then to oversow stylo seed. Fire removes standing and fallen litter, provides a favourable seed bed and reduces competition from the native grasses.

Fire can also be used to manipulate the balance of grass and stylo within the pasture. Cattle select green grass during the growing season and under heavy stocking, grass plants can decline and stylo becomes dominant. Although cattle still grow well on the higher quality legume leaf, the stemmy stylos provide poor ground cover. Resting and burning can restore the balance between grass and legume; the stems of mature shrubby stylos are killed but the legume recovers from seed in the soil or by sprouting from near ground level.

Burn just before or after the first storm rains.

Resting and burning removes the top growth of shrubby stylo. Grass–legume balance is restored as grasses recover, stylo comes back by resprouting from the base or from seedlings.

Under heavy grazing, native grasses decline and shrubby stylo can become dominant.

Rotational burning in perennial tall-grass pastures

Fire can improve the distribution of grazing animals and prevent or minimise the formation of patches prone to degradation.

Cattle can overgraze patches on recently burnt country during the wet season but they can be enticed away from these patches in the next wet season by burning another portion of the paddock. Animal production is improved as they have access to higher quality feed. Using such a fire regime and light grazing (average of 12% utilisation) has been successful in experimental paddocks near Katherine for 18 years.

The trends in perennial grass content of these pastures and animal production are shown in the figure below. They demonstrate that both perennial grasses and animal production can be maintained in these systems with this management regime. Although not yet fully tested, fire could also prove useful for manipulating grazing distribution in large paddocks containing a number of plant communities.

by Andrew Ash

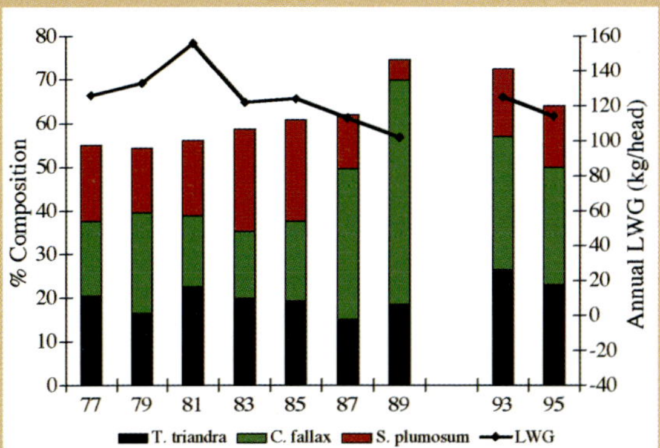

Perennial grass composition and animal weight gain (LWG)

Variation in the perennial grass composition and animal production from a lightly grazed (1 AE/16 ha) and biennially burnt tall-grass pasture at Manbulloo near Katherine, NT over an 18-year period

Rosewood Station, north-west NT

Rosewood Station lies 100 km south-west of Timber Creek, in north-western Northern Territory. It runs a total of 27,000 head (when fully stocked) on 3000 sq. km. Rainfall averages 630 mm a year. Mitchell and Flinders grass on black soils provide more valuable grazing than the taller species and spinifex on the hilly red soils. Fire is used on the property to reduce the risk of wildfire, manage grazing behavior and to assist with pasture regeneration.

To the east, Rosewood borders Aboriginal land, much of which is very lightly stocked; the resulting high loads of fuel can create a major wildfire risk. To reduce the risk of fires crossing the boundary, manager Doug Struber has several burning strategies.

Fuel reduction strategies

Rosewood has over 700 km of roads that make excellent firebreaks when graded with two cuts; they can be used to back-burn off in the early wet.

Fuel reduction burning begins in early December and finishes in April. Helicopter and 4WD are used for opportunistic burning of areas where heavy fuel loads create a risk.

Burning the red hill country promotes fresh growth and encourages cattle to move out to the hills away from the bores in the wet season. Burning, then grazing removes the threat of wildfire for about another three years; it also allows the pasture around the major watering points to regenerate.

Doug does not burn the black soil country as he is not sure how the species there respond.

Legume establishment

Fire is also used on Rosewood to establish seca stylo (*Stylosanthes scabra*) on the red soil areas. Burning before stylo seed is spread from the air creates a bed of ash, allows the seed to reach the soil surface and reduces competition from other grasses (see p. 58).

Effective fire and pasture management on Rosewood Station relies on the identification of areas of high fuel load and on good knowledge of the country. Boundaries, roads and fence lines are well maintained for access to country and firebreaks, and good relations with neighbours are maintained to ensure that wildfires are not a major threat to land and livestock.

by Andrea Johnson

Using fire to manage trees and shrubs

How have tree and shrub populations changed?

The balance between trees, shrubs and grass throughout northern savanna systems has probably always been changing, mostly in relation to variable fire regimes and rainfall. These changes rarely proceed at a steady rate. For example, in areas with erratic rainfall, trees and shrubs grow from seed in episodic bursts, and widespread 'dieback' of mature trees due to drought occurred in parts of the NT in the 1930s and 1960s and in north Queensland in 1990s.

Dramatic changes in tree cover and pasture condition occurred on calcareous red soil pastures at Kidman Springs, VRD between 1973 and 2000 as a result of no burning and reduced grazing pressure.

Why have trees and shrubs increased?

Since European settlement, tree and shrub increases and thickening has been recorded throughout many open grasslands and woodlands across northern Australia. Reduced incidence and intensity of burning in many areas over the last 50 years is a likely contributing factor, a result of deliberate fire exclusion or lower fuel loads caused by grazing (see p. 38).

Increases in species of *Eucalyptus* and *Acacia* in forests and woodlands, as well as shrubby undergrowth species such as *Carissa ovata* (currant bush), *Carissa lanceolata* (conkerberry) and *Hakea arborescens* (common hakea) appear to be widespread. Small tree and shrub species such as *Terminalia volucris* (rosewood), *Lysiphyllum cunninghamii* (bauhinia) and *Excocaria parviflora* (gutta-percha) have also increased on more fertile cracking clay pastures, while *Melaleuca* species are invading grassland swamps in north Queensland.

Figure 5.1 Effect of woody vegetation on pasture growth

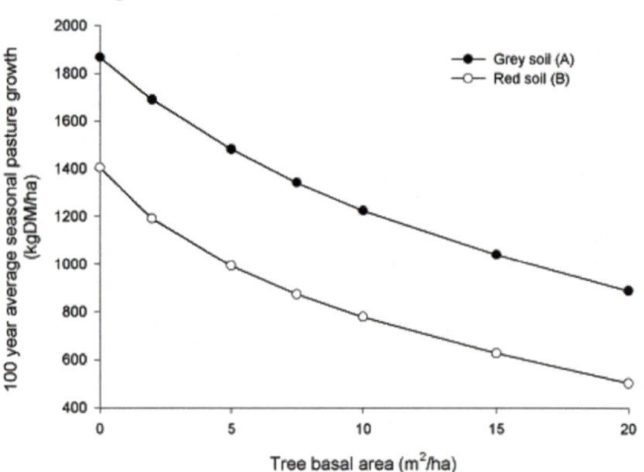

The influence of woody vegetation (tree basal area) on modelled seasonal pasture production for the southern Victoria River District, Northern Territory

Shrubs invading grassland

What are the impacts of increasing trees and shrubs?

Increases in tree and shrub populations has major implications for:

Pastoral industries

Increasing woody vegetation can reduce the forage supply for livestock (Figure 5.1), reduce the access to water, make mustering more difficult and harbour pest animals such as feral pigs.

Conservation

Shifts in the amount of woody vegetation alter the proportions of different habitats.

Tourism

More or fewer trees and shrubs may alter the aesthetic value of a landscape.

What is the role of fire?

Fire is the most effective tool to manipulate the amount of woody vegetation present in uncleared savanna landscapes. The objective of prescribed burning should be to manage, rather than control, the composition (species mix) and structure (density and height) of trees and shrubs. To do this we need to understand how the communities respond to particular fire regimes.

The impact of a single fire on plants is determined by their response to burning (which depends on the species concerned) and the amount of plant fire damage (which is determined by fire intensity and plant height). The effect of whole fire regimes on plant communities is also influenced by the frequency and intensity (timing, season and fuel loads) of burning.

How do trees and shrubs respond to burning?

For trees and shrubs that survive burning by resprouting, such as most eucalypts, fire should be used to manage tree height and canopy size rather than density. If growing points on branches are not damaged sufficiently by fire, leaves will resprout. With increasing fire damage, larger branches and stems are killed (top-kill) and plants survive by resprouting from the protected buds near or under the soil surface.

For species that are killed by fire and regenerate from seed (obligate seeders), fire should be used to reduce plant density. For some fire-sensitive communities such as sandstone heath, if seedlings are reburnt before to maturity and seed production, entire populations may be decimated. For other communities such as lancewood (*A. shirleyi*), large stores of hard seed in the soil enable continued seedling establishment over several years.

Seedlings killed, stems of shrubs dead (top-kill), some trees defoliated after an intense fire

*Following death of stems and branches from intense fires some shrubs such as rosewood (*Terminalia volucris*) resprout from the base.*

Management of species that regenerate profusely following burning, whether they be obligate seeders such as mulga (*A. aneura*) or 'turpentine' (*A. lysiphloia*), or vegetative reprouters such as 'needlebush' (*A. farnesiana*), relies on periodic burning over a number of years.

Why is tree height important?

For fire-sensitive obligate seeders, tree height is not particularly critical for determining plant death. For 'resprouters' however, as plants become taller, they generally become more resistant to fire and require higher fuel load and cover levels to cause death or sufficient plant damage to result in top kill (Figures 5.2 and 5.3). Where the dominant increasing tree and shrubs species are 'resprouters', plant growth should be periodically interrupted by burning before heights exceed 2 m. Once plants exceed this height, top kill and subsequent control is difficult to achieve with manageable fires. Only very high fuel loads and fire intensity will have any effect.

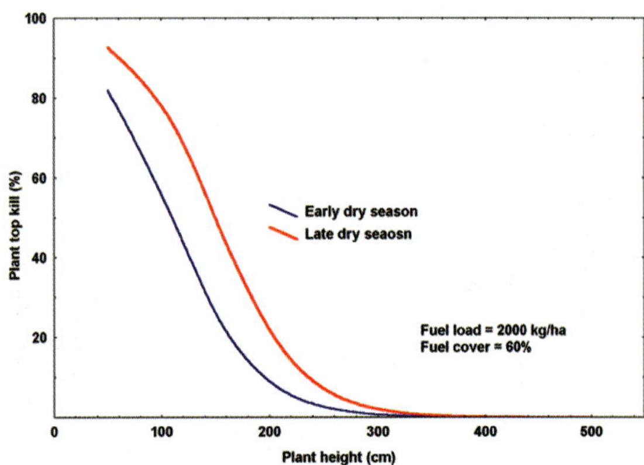

Many acacias are susceptible to fire but regenerate from seed.

Figure 5.2 Plant mortality and height

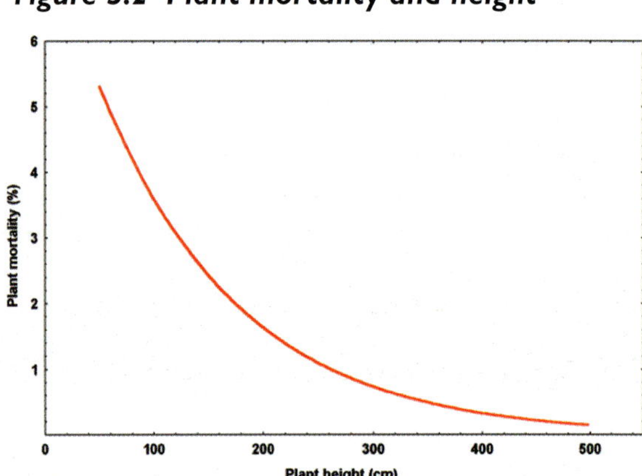

Figure 5.3 Plant top kill and height

As 'resprouting' species such as rosewood increase in height, plant mortality (Fig. 5.2) and top kill (Fig. 5.3) falls rapidly. Late dry season fires cause higher levels of plant top kill than early season fires.

Fire management on Trafalgar Station, Charters Towers

Trafalgar Station near Charters Towers is a 32,000 hectare property with a carrying capacity of about 3200 adult equivalents.

Pasture management and fire

I aim to 'spell' 15–20% of the property each year to allow the grass to strengthen and drop seed; the build-up of pasture gives us an option to burn.

If the season is poor, I may keep the grass for drought feed or burn later. However, paddocks that have been locked up after clearing virgin scrub are always burnt and allowed a full growing season before being grazed.

Fire management

The pasture is usually burned in October–November when the weather is hot, slightly humid and the winds are fairly light, preferably from the north. Hot, dry winds tend to push the fire along too quickly, burning the grass but not the weeds or fallen timber.

The conditions must be right before I burn. Having gone to the trouble of planning the burn, spelling the paddock, spending money on fire preparations, getting experienced people in the right numbers and the right equipment for any emergencies, I'm not going to waste the fuel and the benefit by burning in the wrong conditions—an overcast sky, the winds in the wrong quarter or winds too light.

After the fire, stock should be excluded until ground cover is well established, preferably after good rain. Our standard practice at Trafalgar is to sow stylos and grasses into the ash bed 3–4 weeks after the fire and before the first rains.

Fire for woody control

Fire is very effective against some weeds.

Currant bush (*Carissa ovata*) burns well once started, even without much grass fuel. Although the adult plant may not be killed it is controlled so that grass will re-establish.

Dense and light infestations of rubber vine in the paddock are effectively controlled. With a heavy grass fuel load, a light wind and slightly moist ground and atmosphere, fire will kill up to 80% of rubber vine plants, especially if they are small. It can carry a fire without additional fuel if conditions are extremely hot and the sap very active in the plant.

Fire plays a significant part in the control of eucalypt seedlings or suckers in open forest when it is used regularly. But in cleared country after the initial burn, burning in later years is a waste of grass as the suckers tend to regrow within two weeks of the fire.

Cost of burning

Fire is often quoted as a cheap option for pasture and land management. The direct costs of setting up firebreaks, getting the right equipment and personnel, and the actual burning may be quite small on a per hectare basis, but locking up a paddock and burning it is definitely not cheap as production is foregone.

Costs that would arise from not burning include clearing regrowth, controlling weeds or timber, and lost grass growth from competition with woody regrowth.

Good pasture management means stocking conservatively, spelling paddocks and burning for a purpose when it is sensible to do so.

by Roger Landsberg

Reference

Landsberg, R. (1997). The use of fire as a management tool in the semi-arid tropics: a producer's perspective. In *Bushfire '97*. (Eds B. J. McKaige, R. J. Williams and W. M. Waggitt.) CSIRO, Darwin. pp. 221–226.

Fire is used to control rubber vine invasion on Trafalgar Station.

Inspecting the result of an intense prescribed burn at Trafalgar Station

Intense fires from dry fuel give the best control of woody plants.

What is the effect of fire intensity?

Fire intensity influences the level of damage and response of plants to burning. High-intensity fires (>2000 kW/m) are most effective for controlling woody plants.

What fuel loads are required?

The amount and arrangement of grassy fuel will determine fire intensity and its impact on trees and shrubs. Generally, as fuel load and cover increase so does plant top kill following burning.

Suppression of smaller plants (1 m in height) can be achieved with fuel loads of 1500 kg/ha and 60% cover. When plants are over 2 m however, fuel loads between 3000–4000 kg/ha combined with cover levels greater than 70% are required for rates of top kill exceeding 30–40%.

When to burn?

As the dry season progresses, changing fuel and weather conditions promote fires of higher intensity, and increase tree and shrub top kill (Figure 5.3). Plant damage and top kill will be increased by carrying out prescribed burning during the late dry season (August–October). Early dry season fires are less intense, patchy and not as effective. Prevailing fuel and weather conditions prior to burning are a better guide to fire intensity and effectiveness compared to burning season alone due to the variability in burning conditions within and between years.

Apart from higher fuel load and cover, high air temperatures (25–35°C), lower humidity (<30%) and higher wind speed (>5 km/h) will increase fire intensity and burning effectiveness (Table 5.1).

How often to burn?

How often you need to burn depends on your objectives of burning, how tall the plants are, the rate of plant regrowth, how soon they will drop seed and how quickly fuel accumulates.

For 'resprouters', a single intense fire will reduce the amount of woody vegetation present, but individual plants may survive—the amount of shrubs is reduced but not shrub density. Periodic intense fires can prevent new shrubs establishing by killing seedlings and saplings, but too frequent burning with low fuel loads will have little effect while wasting valuable forage for grazing and possibly damaging the pasture plants.

Table 5.1 Burning conditions

Maximum tree/shrub height (cm)	Relative humidity (%)	Wind speed (km/h)	Fuel load (kg/ha)	Fire intensity (kW/m)
50	30	5	2200	1000
100	30	8	2500	1400
150	30	10	3000	2200
200	30	12	3500	3000
300	30	15	4900	4600
50	50	8	2400	1100
100	50	10	2900	1800
150	50	12	3400	2600
200	50	13	4300	3500
300	50	15	5900	5300

Burning conditions (relative humidity, wind speed, fuel load and fire intensity) necessary to achieve an 80% top kill of woody plants with increasing height classes in the Victoria River District and pasture cover of 60%

Low, discontinuous fuel loads generally support low-intensity, patchy fires.

The optimum fire frequency for controlling tree and shrub species increasing on pastoral lands in the semi-arid savannas of the Victoria River District is 5–7 years. Less frequent burning allows the plants to reach over 2 m in height and so become more resistant. In regions with lower rainfall and with higher grazing pressure, longer intervals between burning may be necessary to accumulate enough fuel.

What is the impact of grazing?

Grazing reduces the amount of grassy fuel and its continuity between plants, resulting in less intense fires. Increasing density of shrubs is frequently a sign of higher stocking rates; these pastures may have to be rested to accumulate enough fuel for an effective fire. On the other hand, a lightly grazed and dense pasture will compete with, or prevent the establishment of, seedlings of trees and shrubs.

Figure 5.4 Tree height and fire

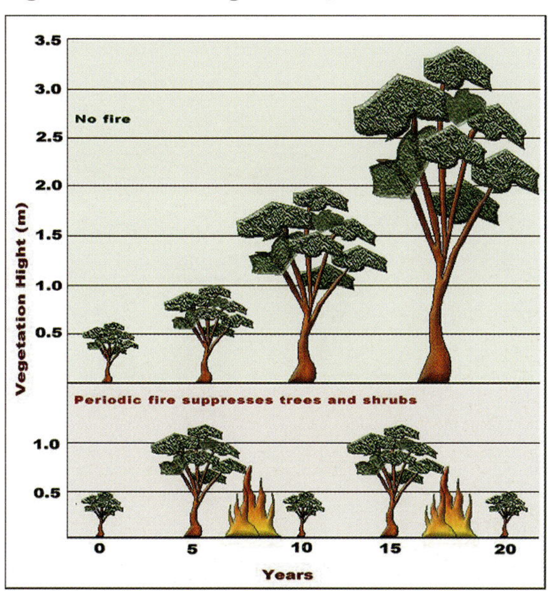

Periodic fire is required to maintain tree and shrub height populations.

Grazing reduces the continuity and bulk of fuel resulting in patchy, low-intensity fires. Heavily grazed country may have to be rested.

Using fire to manage for exotic weeds

Exotic weeds threaten pastoralism, conservation or tourism in many parts of the extensive northern savannas. Fire can be used to manage these exotic weeds, but it demands an ecologically based fire regime rather than a single fire.

Why use fire?

The advantages of fire for managing fire-sensitive weeds in extensive savannas include:

- Many native species, including grasses, are not susceptible.

- It is relatively inexpensive to treat large areas.

- Individual plants do not have to be located.

- Fire can be employed as a preventative measure.

What are the key aspects for weed control?

The biology of a weed species must be understood to design an appropriate burning regime. Key questions are:

- How long before the plant can reproduce?

- How big before the plant becomes resistant to fire?

- How often do the seedlings establish?

- How does the seed spread and how far?

Rubber vine (*Cryptostegia grandiflora*) and one of the mesquite species (*Prosopis pallida*) are relatively vulnerable to fire, but their populations do recover from seed slowly. As the seeds of rubber vine do not stay viable for long on the ground it can be eliminated by killing the adult plants. However, other mesquite species (*P. glandulosa, P. velutina* and *P. juliflora*) and chinee apple (*Ziziphus mauritiana*) are relatively resilient to fire.

Species of intermediate susceptibility to fire include: prickly acacia (*Acacia nilotica*), parkinsonia (*Parkinsonia aculeata*), rubber bush (*Calotropis procera*), giant sensitive plant (*Mimosa pigra*) and bellyache bush (*Jatropha gossypiifolia*).

All species are probably more vulnerable to fire when they are young, especially newly emerged seedlings, although chinee apple, for example, can resprout very early in its life.

Will fire alone be enough?

Fire can never be the total solution to controlling weeds, and has to be integrated with chemical and mechanical methods and biological control, along with quarantine to prevent the weeds being introduced or spread.

As with native plants, some exotic weeds are susceptible to fire, others resilient. Some are easily killed but come back from seed; others can sprout from protected growing points.

What are some of the costs involved?

Although fire is relatively inexpensive, there are still economic costs—including those associated with destocking to prepare an area for burning or to allow it to recover following fire, with constructing or maintaining firebreaks and in conducting the burn itself.

Fire may have an important role in the control of bellyache bush (top), mimosa (centre) and parkinsonia (bottom).

Managing rubber vine with fire

Rubber vine (*Cryptostegia grandiflora*) is an invasive weed that is choking out many woodlands and riverside areas throughout northern Queensland. Not only does it create dense thickets, removing pastoral country from production, it is a refuge for feral pigs and removes ground cover promoting erosion along creeks and rivers. Rubber vine is relatively fire sensitive provided the stem base of each plant is heated. Burning in the late dry season will yield high-intensity fires that may kill most juvenile plants and 50–70% of adult plants. Fires of this intensity need 3–4 t/ha of grass fuel.

Burning on hot days during dry spells in the wet season can also be effective. This method has been used in riverside zones and relies on having significant amounts of litter (e.g. eucalypt leaves, branches and similar debris). This litter provides enough fuel to reach the ignition temperature of green rubber vine leaves, leading to an intense fire.

by Tony Grice

Further reading

Grice, A. C. (1997). *Australian Journal of Ecology* **22**: 49–55.

High-intensity fires can kill and open up dense rubber vine infestations.

High fuel loads and high-intensity fires are most effective for rubber vine control.

Rubber vine infestation chokes out river systems and woodlands.

Burning to control mesquite

Fire is a highly effective technique for controlling *Prosopis pallida*, the most widespread mesquite species in Australia. The best kill comes from burning late in the dry season when the plants are stressed and the fires are intense; mature trees, seedlings and seeds lying on the soil surface are susceptible.

Fire can be effective against scattered mesquite where there is plenty of grass or against dense patches of the shrub where there is no grass but a lot of leaf litter on the ground. For scattered infestations, grazing must be controlled to allow sufficient fuel (at least 2000 kg/ha) to build-up. Post-fire grazing management is also important as a good grass cover helps to suppress seedling regrowth.

In most cases, a single fire will not control an infestation completely and should be followed with either another burn or with chemical or mechanical control. Burning every three to five years will generally keep *P. pallida* in check.

by Shane Campbell

Late dry season, high-intensity fires are most effective for mesquite control.

Mortality rates of mesquite (Prosopis pallida) *following burning can be high if conditions are right.*

Grass production is increased under killed mesquite canopies following burning.

Managing for wildfires

Wildfires burn across vast areas of northern Australia every year (see Figure 1.1 p. 1).

These bushfires are very different from those in southern states—fuel loads are much lighter and the trees themselves rarely burn.

Bushfires in the savannas are usually fast-moving grassfires.

Wildfires are not necessarily intense or bad—just unplanned and uncontrolled; however, if extensive and frequent, they can cause damage whether the land is used for pastoralism, conservation, defence or traditional use.

What are the impacts of wildfires?

The economic and ecological costs of wildfires include:

- cost of firefighting and prevention—grading firebreaks, labour, fuel

- damage to fences, buildings, vehicles

- loss of forage for livestock

- disrupted livestock management—mustering and lost production

- reduced biodiversity and habitat damage

- loss of food supply for native fauna and traditional occupiers

- atmospheric pollution—smoke and greenhouse gases

These costs may not be sufficient to justify fighting an extensive fire once it has started, making the only realistic defence some form of fire management such as pre-emptive hazard reduction.

In the far north, distances and the size of wildfires can be enormous—measured in thousands of square kilometres burnt rather than hectares—they burn over weeks or even months in remote areas and often break up to create several fronts burning at the same time. They may be extinguished only by natural breaks, the weather or by burning into previously burnt areas. One or two graders, a small number of station workers and a couple of small firefighting units often control huge fire fronts, but the operations can continue for days or weeks.

Resources, including manpower, are usually scarce and communications in remote areas have only recently improved with the development of satellite technology. Fires are now often detected by satellite, while light aircraft are used to locate and quickly assess them. Aircraft are also used to assess the success of control operations.

How do wildfires start?

Lightning at the start of the wet season may once have been the main cause of fires but human activities—arson, prescribed fires that escape, military activities and carelessness—are now more common causes of ignition. Careless activity can include sparks from welding, and accidents can include sparks arising when a grader blade strikes a rock.

Where are wildfires most likely?

Wildfires are more likely where there is more grass growing and less use made of it.

The risk can depend on:

Vegetation type

In high-rainfall areas, tall annual sorghums and introduced species such as gamba grass can produce massive loads of fuel each year and this dries or 'cures' rapidly as the annual life cycle ends, creating a serious fire hazard during the dry season.

Wildfires can burn across vast areas of savanna.

Fast-moving grassfires of the savannas are much less intense than the bushfires of southern states.

Seasonal rainfall

Above-average rainfall results in rapid accumulation of herbage and fuel, even in normally arid areas. This caused the massive wildfires in central Australia during the 1970s.

Land use

Wildfire risks are greater on lands that have low–intensity management, are sparsely settled and ungrazed. Human ignition sources also increase the incidence of wildfires along ungrazed roadsides, around settlements and where the army is 'live-firing' on defence lands.

Grazing intensity

Heavier grazing pressure decreases the amount and continuity of fuel and reduces the risk of fire spread and intensity. On pastures of low nutritive value, such as annual *Sorghum* and tall-grass pastures, stocking rates are low and little grass is eaten leaving high fuel loads.

Tools to assist in wildfire management

Tools that can assist in wildfire management include:

- fire hotspot maps

- fire scar history maps

- photo-standards of fuel loads

- fuel curing maps and photo-standards

- fire risk maps

- fire danger index (McArthur Grassland Fire Danger Meter)

- meteorological data and general information on fire weather patterns

Descriptions of these tools and their use in fire management and planning are explained in greater detail in Chapter 7.

How can wildfires be reduced?

It is usually not practical or feasible to eliminate wildfires but the risk can be reduced by developing strategic fire management plans.

Figure 5.5 VRD Station pasture growth

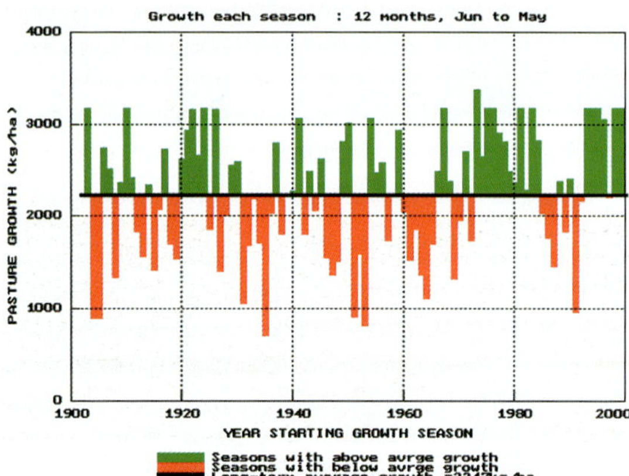

Heavy herbage growth in years of above-average rainfall can create a fire hazard, even in normally arid areas.

The risk of wildfires can be reduced with strategic fire management plans.

Fire management at Ban Ban Springs

My 14 years of experience at Ban Ban Springs about 130 km south-east of Darwin illustrate one way that a range of fire-related issues can be dealt with practically.

Ban Ban Springs receives an average 1200 mm rainfall. The topography varies between high, steep ridges in the east and west and the extensive flood plain along the Margaret and McKinlay rivers. The open savanna woodlands have a typical Top End overstorey of trees and shrubs over kangaroo grass (*Themeda triandra*), but with annual and perennial sorghums (*Sorghum* spp.). Cattle graze these native pastures at stocking rates of about one beast to 10 ha in relatively large paddocks (3000–15,000 ha).

Our fire management

Our burning regime revolves around extensive burning during the early to mid–dry season, coupled with selective wet season burning to control tree saplings and the *Sorghum* grasses. However, if stocking rates are too heavy (over 1:10 ha), repeated dry season burning can rapidly deplete pastures— within three seasons. High grazing pressure also favours growth of shrubs and trees as well as introduced weeds.

Timing of fire application

Ideally we burn about one-third of the grazed part of the property each year in the late wet season or early dry season (March–May) giving a three-year cycle.

Practically, it has been possible to burn about half of each paddock, although wildfires can upset the intended burning regime.

Burning on Ban Ban can be opportunistic and generally takes place before the ground has dried sufficiently to grade firebreaks. We make use of the Bushfires Council helicopter to drop incendiaries when access to the country is still limited. An aerial drip torch is used to ensure that secure burnt lines are established in areas that are too wet or green for effective incendiary ignition.

Grazing the breeding herd on this combination of burnt and unburnt pastures allows us to wean calves earlier in the dry season so that the breeding cows maintain better condition later into the season and so conceive earlier.

Fire used in this way also helps to even out the grazing pressure as cattle prefer to graze young, short, green pasture that is provided by burning and, to a lesser degree, by grazing. Grazing pressure is shifted dramatically within days to the burnt areas where it is common to find 90% or more of the stock. This provides another advantage; our stock are concentrated and so easier to muster.

Late dry season fire is an effective means of controlling woody weeds, but the paddock should be lightly stocked (one beast to 30 ha) to build up grassy fuel loads for the high-intensity fire needed. These kinds of burn should take place after rainfall of at least 100 mm or at the onset of the full wet season to reduce the risk of total pasture depletion and to encourage rapid pasture regrowth.

Fire history maps from the Bushfires Council help us with planning and monitoring. Knowing where the burnt areas are saves time and resources when combating wildfire later in the dry season.

by Tom Starr

Ban Ban Springs is about 130 km south-east of Darwin.

Tom Starr (centre)—former manager, Ban Ban Springs

What strategic options are useful?

Several strategies can be used to reduce the risk of wildfires:

Hazard reduction

Strategic prescribed burning can reduce the amount and continuity of grassy fuel. Burning to reduce fuel during the early dry season—before the fuel is fully cured—results in low-intensity, patchy burns and forms a mosaic of fuel loads.

Firebreaks

Firebreaks are rarely effective at stopping a headfire (see p. 88). Their main roles are to provide a safe, clear line from which to conduct back-burning operations, and to protect strategic boundaries and areas of importance.

Strategic grazing

Manipulating the number and location of livestock in strategic areas can influence the amount and distribution of fuel.

How can fuel hazards be reduced?

Fuel loads can be reduced by:

- burning the roadsides

- rotational burning within paddocks

- mosaic burning within landscapes

These burning strategies are represented diagrammatically in Figure 5.6. Rotational and mosaic burning strategies can be used to meet multiple fire management objectives in addition to wildfire hazard reduction. On pastoral lands, they will also have a significant impact on how cattle grazing pressure and fuel continuity is distributed around the paddock.

Extensive areas of land are most effectively treated using a combination of aerial and ground ignition. Aerial ignition can be carried out along pre-determined strategic fire lines or in a regular grid pattern using point ignitions, while lighting up from the ground continuously around a perimeter can increase the spread and intensity of a fire and can result in a more complete burn.

When is the best time to burn to reduce fuel?

Fires should be planned to burn off sufficient fuel to minimise the risk of wildfires later in the dry season but without them becoming a problem themselves. They have to be timed to coincide with a degree of curing and the right weather conditions.

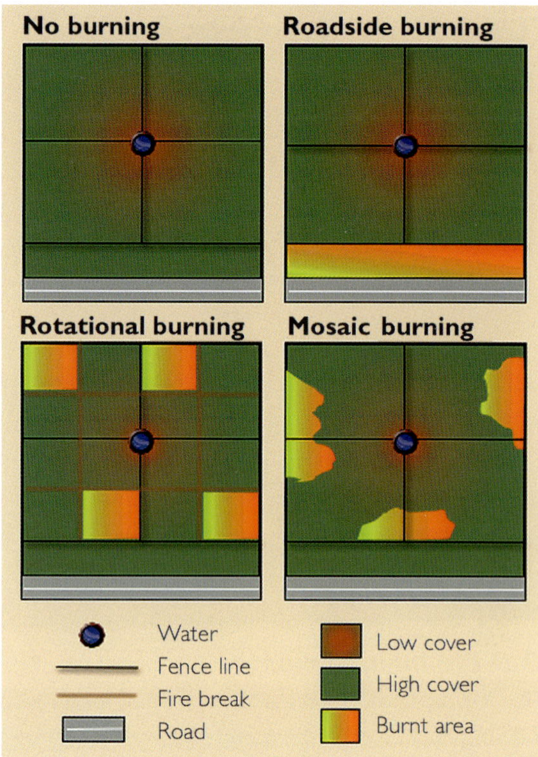

Figure 5.6 Burning strategies

Roadside, rotational and mosaic burning are effective at reducing fuel loads and limiting fires entering and spreading throughout the landscape. Rotational and mosaic strategies are also effective for achieving a range of other fire management objectives.

Firebreaks provide a safe, clear line from which to conduct back-burning operations, and to protect strategic boundaries and areas of importance.

Fuel loads can be reduced by burning the roadside verge.

Season

Fuel curing state, flammability and weather conditions vary during the year and between years. Although fuel reduction burning is best carried out in the late wet or early dry season, it is difficult to decide exactly when as there is usually considerable variation in curing across the landscape.

Is it worth trying to put out the fire?

The economics of fire management in the north are generally not well understood.

The issues include:

- What is the cost of putting out the fire?

- What is the damage or benefit from the fire?

- What ecological values are at risk?

Control measures are usually undertaken only where bushfires threaten productive pastoral areas or assets such as buildings or infrastructure. Existing bushfires are controlled—if necessary—by back-burning from natural or man-made firebreaks.

Smaller fires in closely settled areas usually attract a concerted response because of the higher populations and the threat to buildings and assets.

Huge fires covering thousands of square kilometres in the remote areas may be left to burn their natural course because they threaten no financial assets although they could have significant environmental effects.

Areas are so large and resources so limited that fire risk is best managed by reducing the potential hazard at a time when burning can be controlled.

Strategic prescribed burning can reduce the amount and continuity of fuel.

Wildfire management on Elsey Station

Max Gorringe manages Elsey Station, 5300 sq. km of savanna woodland owned by the Mangarrayi Aboriginal Land Trust. Wildfires sweep through Elsey almost every year threatening feed reserves for the 13,000 cattle, station infrastructure and important habitats. Most come from the eastern side of the station between August and October. Many start in parking bays along the Roper Highway and are caused by careless use of fire by people passing through to Roper Bar; others start along the Stuart Highway and old Elsey cemetery.

What does firefighting cost us?

Costs for Elsey Station in 1999 were astronomical as two-thirds of the station were burnt out between August and October and we lost a Toyota. Total costs arising from wildfires exceeded $89,000 and included grading firebreaks, aircraft operation, labour and mustering costs.

What does wildfire prevention cost?

The costs attributed to wildfire prevention total around $26,000 each year. This includes costs associated with grading firebreaks, aerial and ground burning from boundaries and roadsides, and patch burning within paddocks.

Fire management

We burn Aerial Control Burn (ACB) lines in the early to mid-dry season every year to reduce fuel along and within boundaries, and to reduce the risk of late dry season fires moving onto the station.

We employ a grader driver full-time throughout the dry season, split between grading firebreaks and maintaining and upgrading fence lines especially in the bush paddocks south of the Roper Highway.

Strategic and controlled burning is carried out throughout the year to reduce fuel loads and reduce woody tree and shrub growth. We burn along the road and boundaries from a vehicle when fuel conditions are right for sustaining small, patchy fires, and opportunistically from the air with capsules each wet season to remove old, dry growth within paddocks.

Preventitive back-burns are carried out along graded fence lines and breaks when time allows. This type of burning requires a continuous line of fire to penetrate the paddock to create a wide enough break to reduce spread of wildfires over the breaks later in the year. This more intensive, controlled burning requires Toyotas, slip-on units and drip torches.

by Max Gorringe and Andrea Johnson

Max Gorringe has a strong fire management plan on Elsey.

Wet land habitat on Elsey

Using fire to manage for biodiversity

A unique and complex array of habitats, plants and animals exist throughout northern Australian savannas. Maintaining this biodiversity is critical if these lands are to remain relatively healthy. The retention of biodiversity relies not only on managing conservation reserves, but also on managing the extensive tracts of 'off reserve' pastoral, Aboriginal and defence lands. Many of the fire management issues identified previously in this chapter relevant to traditional Aboriginal and pastoral land-use objectives apply similarly to the use of fire for managing biodiversity (e.g. developing habitat patchiness, weed control, and wildfire management).

What role can fire play?

Our knowledge of the complex impacts of fire on habitat, plant and animal species is less than perfect. Fire has a special role in maintaining biodiversity throughout the landscape via the creation of habitat diversity, both through the creation of patches (i.e. spatially) and over time (temporally). This involves applying a variety of fire regimes with different characteristics—burning at different seasons, frequencies, intensities and patch sizes.

What are the main objectives of fire management for biodiversity?

The objectives for biodiversity are to:

- maintain and promote the diversity of habitats, plant and animal species

- protect fire-sensitive vegetation communities and wildlife habitat

On conservation lands, the protection of infrastructure, high-use areas and the maintenance of asthetically pleasing landscapes are also important objectives.

How do fire regimes maintain habitat diversity?

As described in the previous chapter, habitats and their species can vary greatly in their susceptibility to burning. For example, within any one region in the savannas there exists a mosaic of habitats ranging from those which are relatively fire-resilient (e.g. savanna woodlands and grasslands) through to typically more restricted fire-sensitive types, including riparian communities, vine thickets, and stands of cypress pine and *Acacia* scrubs (e.g. lancewood, *Acacia shirleyi*).

Keeping such relatively fire-sensitive habitats in the landscape very much depends on the severity of fire regimes.

Are animals impacted by fire?

The abundance of invertebrate, bird, reptile and mammal species is partly determined by their adaptation to specific fire regimes. For example, in semi-arid grasslands certain species of ants, birds and reptiles are associated with recently burnt areas while others occur mainly in areas that have been long unburnt (Table 4.3 p. 43).

The survival of some species also relies on their access to a range of habitat types within their home range. This includes having access to areas with different burning histories—from recently burnt to long unburnt. Seed-eating birds such as the Gouldian finch and reptiles such as the frillneck lizard rely on long unburnt areas for protection and refuge but use recently burnt areas for gathering food.

What is the impact of changing fire regimes?

Fire regimes have been evolving in northern Australia over millions of years—from the time that former extensive tracts of rainforest began to give way to the savannas under an increasingly seasonally arid climate. Further, such aridity has not developed steadily or progressively—there have been at least 20 major climatic cycles over the past couple of million years alone, ranging from relatively short wet periods (such as the past 10,000 years) through to extended harsher periods where dune fields have covered as much as two-thirds of the continent.

Over the past 50,000 years, regional fire regimes have been governed principally by people. As such, habitats and component species have had to adapt to dynamically changing climates and varying fire regimes for a considerable period.

Before the arrival of people, fires probably occurred mostly in the build-up period at the start of the wet season, ignited by lightning strikes. Since the arrival of people, fires are likely to have been lit throughout the dry season. In applying fire as a management tool today, it is necessary for us all to appreciate the potent roles that fire regimes can play in enhancing, maintaining or degrading regional biodiversity.

What are some of the key decisions relating to fire management?

- What mix of fire regimes (frequency, season and extent) should be aimed for in the landscape?

- How large should individual patches (fires) be?

- How should the patches be distributed relative to each other?

Is there a correct fire frequency?

Management should aim to achieve a patchwork of fire ages, from frequently burnt to unburnt areas, throughout the landscape. Fire regimes that are repeatedly and uniformly imposed over large areas (including the absence of fire) will disadvantage some species.

Annual, late dry season fires currently occur over large areas of the Kimberley, the Top End of the NT and western Cape York. Such extensive, uniform fire regimes can be destructive to fire-sensitive species, including fauna which are dependent on structurally diverse vegetation for seasonally critical food resources or lodging (e.g. hollow logs). A further casualty concerns those fire-sensitive plant species that require several years between fires for seedlings to mature and set seed to ensure their survival. High-frequency fire regimes also affect the floristic diversity and structure of dominant forest and woodland species generally, creating communities dominated by mature adults and suppressed understorey resprouts, whilst being devoid of any mid-storey (Table 4.2 p. 36).

A balance between too much and too little fire is needed. Reduced burning in wet-sclerophyll forest habitats adjacent to rainforest in north-eastern Queensland has resulted in invasion by rainforest species and associated reduction of a fire-carrying grassy understorey. Absence of, or reduced, landscape-scale burning in more closely settled pastoral areas of western Queensland and on productive Mitchell grasslands across the north has, in some areas, aided invasion of exotic weed species.

Why is season and timing of burning important?

Seasonality and timing of burns are critical factors, not least because they provide important means by which fire intensity and spread can be effectively managed. Low-intensity, patchy burns lit early in the dry season, at night, or in the wet season, are useful for implementing protective burns on the margins of, or even within, fire-sensitive communities. For example, patchy burning in heath communities which contain obligate-seeder shrub species, will develop a mosaic of different vegetation age classes. By contrast, a uniform, extensive fire exposes the next generation of seedlings to risk from being burnt in a subsequent fire before they have attained reproductive maturity.

Alternatively, intense fires may be required to control certain woody species. In the case of controlling tea-tree (*Melaleuca*) regeneration on flood plain grasslands, the best time for using an intense fire appears to be after the first rains of the wet season. Given available soil moisture, burning at this time assists the rapid regeneration of grasses and sedges which can then compete with, and thus reduce the vigour of, regrowth from tea-tree suckers.

A further critical issue concerning the timing of burning is to ensure that such burning does not interfere with the breeding cycles of native animals. For example, various ground-nesting (e.g. partridge pigeon, whistling duck) or low foliage-nesting (e.g. masked finches) birds are vulnerable to fires lit very early in the dry season. Similarly, towards the end of the dry season young quolls are at risk given that mothers and their young use hollow logs and other ground-level lodging through the day.

As Melaleuca *scrub (paperbarks) invades flooded grasslands in north Queensland and increases in size, grassy fuel loads decline and trees become resistant to subsequent control with fire.*

Kakadu National Park fire management study

The complexity of managing fire in savanna landscapes is illustrated by the fire management program in the 20,000 sq. km World Heritage Kakadu National Park. Kakadu contains a diversity of fire-sensitive species and habitats, a major town, two extensive and active mining leases, and a number of Aboriginal outstations, ranger stations and tourist destinations.

This complexity is reflected further in the variety of fire management objectives as set out in the Plan of Management. The aims are to:

- promote traditional ways of burning within the park

- involve Bininj/Mungguy (i.e. the traditional owners) in planning and implementing fire management

- protect life and property within and adjacent to the park

- restrict fire from spreading so that it does not enter or leave the park

- maintain biodiversity through effective fire management of species and habitats

In practical terms, fire management is undertaken both by traditional owners and park staff, mostly in the early to mid-dry season period (typically May–July) when fires tend to be small, patchy, of low intensity and typically go out at night under cool, dewy conditions. Burning is undertaken as the country dries out, starting in upland areas early in the dry season, increasingly in moister areas (e.g. creek lines and flood plains) with the progression of the dry season. Fires are lit off tracks and roads, and also with aerial incendiaries using helicopters, especially in more remote locations. More than half of the extensive lowland savannas are burnt on average each year.

Importantly, fire management in Kakadu also concerns much consultation between traditional owners, park staff, park communities, industry, neighbouring properties and the Bushfires Council of the NT. This consultation starts each year before the burning program commences and continues right through to the start of the next wet season. There is plenty of room for things to go wrong.

An active management principle therefore is to break the country up in the early part of the year with a mosaic of patchy fires. Such a mosaic has a number of practical benefits:

- It can halt, or at least reduce, the passage of potentially extensive, typically intense fires which can readily get away in the late dry season under highly flammable climatic and fuel conditions.

- It will foster the development of habitat diversity, providing a matrix of recently burnt, through to long unburnt patches, as required by different groups of plants and animals.

Since 1979, when the first stage of the park was declared, the fire management record has shown some substantial successes as well as ongoing challenges. Amongst the successes: a recent analysis of the fire history of the park developed from interpretation of satellite imagery has shown that the burning program has increasingly been concentrated in the early to mid-dry season in line with traditional practice; that the sizes of individual fires have become markedly smaller and that, over the past decade or so, few uncontrolled fires have entered or left the park. Major challenges still facing the park include: ecologically unsustainable, high fire frequencies in some habitats (including lowland rainforests and fire-sensitive sandstone heaths) and limited opportunities for traditional owners to be actively employed in park fire management programs.

by Jeremy Russell-Smith

Further reading

Russell-Smith, J. et al. (1997). *Journal of Applied Ecology* **34:** 748–766.

Kakadu presents many challenges for fire management, including the protection of fire-sensitive communities of the rugged sandstone Arnhem Plateau.

How can we create patchiness through burning?

Given the importance of fire patchiness in order to create habitat diversity, a key question is how to develop and maintain such patchiness across the landscape. An obvious first point is that to achieve fine-grained patchiness of tens of hectares, individual fires must be small in size. Such fine-scale patchiness, while evidently achievable under intensive land use and management (see the example of traditional Aboriginal burning given at the start of this chapter), is not readily achievable over much of northern Australia today. This is due to typically very sparse human settlement, limited time and other resources and, just as significantly, a diversity of current land-use objectives.

In practice therefore, considerable effort and resources are typically required to 'break up' very large areas by burning onto natural barriers such as creek lines and other drainage areas, or onto (or away from) built structures such as roads, tracks and graded firebreaks. The use of natural firebreaks is ecologically better, and certainly more economic, than reliance on built firebreaks—especially given that tracks provide ready corridors for weed invasion and are prone to erosion if not properly built and maintained. However, it is important to emphasise that flexibility is an essential component of all fire management practice and that a mix of different approaches may often be required.

Practical management tools for developing patchiness include the use of progressive burning and point ignitions.

Progressive burning refers to the undertaking of burning from the end of the wet season, typically in areas of higher ground where grasses cure first, and then proceeding to burn down-slope throughout the year as these dry out. Consequences are that individual fires can be small but of varying intensity (depending on, for example, wind speed), and relatively easily controlled given that fires can be directed onto other burnt areas or areas of uncured fuel just by using wind direction. This was the basis of much traditional Aboriginal burning practice.

Point ignitions (i.e. single ignition from a match) are much better for developing patchiness than line ignitions (e.g. burning off from a firebreak over 1 km using a drip torch). Given a more-or-less constant wind direction, point ignitions typically develop into elliptically-shaped fires. Maximum fire intensities occur at the moving fire front, but the laterally spreading flanking fires are of much lower intensity (Figure 3.6 p. 23).

To create a mosaic of fire intensities, prescribed burning should be undertaken progressively.

To promote habitat diversity, uniform removal of ground cover should be avoided in favour of a more patchy distribution of fuels and fire.

Once dusk falls during the late wet and early dry season, fires generally decrease in intensity, often going out as temperatures drop and humidity increases.

Exotic grasses such as para grass (Urochloa mutica *formerly* Brachiaria mutica) *can present fire management problems when uncontrolled. Here a dense, highly flammable fuel load of para grass abuts a rainforest on a coastal flood plain.*

Fire-sensitive cypress pine seedlings survive low-intensity burns.

Concluding remarks

In finishing this chapter it is important to emphasise two major points:

Monitoring

Monitoring is essential for evaluating the effect of current fire regimes and planning future prescribed burning activities. As described in Chapter 7, a range of satellite and ground-based monitoring tools are available to assist fire management planning decisions.

Communication

Differences in attitude to the use of fire often exist between neighbouring land managers. Such differences can be exacerbated when different, if not multiple, land-use objectives (e.g. pastoralism, tourism, conservation, defence, indigenous and mining exploration), are pursued on neighbouring properties.

Effective fire management requires good communication to explain particular management requirements, and to ensure that cooperative arrangements can be developed.

Early burning around fire-sensitive communities like this floodplain-fringing rainforest provide protection from fires later in the year.

Further reading

Andersen, A. N. (1999). Fire management in northern Australia: beyond command-and-control. *Australian Biologist* **12:** 63–70.

Braithwaite, R. W. (1991). Aboriginal fire regimes of Monsoonal Australia in the (19th) century. *Search* **22(7):** 247–249.

Gill, A. M., Groves, R. H. and Noble, I. R. (1981). *Fire in the Australian Biota.* Aust. Acad. Science, Canberra. pp. 425–439.

Gill, A. M., Hoare, J. R. L. and Cheney, N. P. (1990). Fires and their effects in the wet/dry tropics of Australia. In *Fire in the tropical biota (ecological studies 84). Ecological process and global change.* (Ed. J. G. Goldammer.) Springer-Verlag, Berlin. pp. 159–178.

Grice A. J. and Slatter, S. M. (1997). *Fire in the management of northern Australian pastoral lands.* Proceedings of a Workshop held in Townsville, Queensland, Australia, April (1996). Occasional Publication No. 8. Tropical Grassland Society of Australia, Brisbane.

Grice, A. J. and Brown J. R. (1996). *Fire and the population ecology of invasive shrubs in the tropical woodlands.* Proceedings of the Nicholson Centenary Meeting, Frontiers of Population Ecology. (1995) Canberra. pp. 589–97.

Hodgkinson, K. C., Harrington, G. N., Griffin, G. F., Noble, J. C. and Young, M. D. (1984). Management of vegetation with fire. In *Management of Australia's Rangelands.* (Eds G. N. Harrington, A. D. Wilson and M. D. Young.) CSIRO, Melbourne. pp. 141–156.

Luke, R. H. and McAthur, A. G. (1978). *Bushfires in Australia.* Aust. Govt. Publ. Serv., Canberra.

Paton, C. J. and Rickert, K. G. (1989). Burning, then resting reduces wiregrass (*Aristida* spp.) in black speargrass pastures. *Tropical Grasslands* **23(4):** 21–218.

Pyne, S. J. (1991). *Burning Bush: A Fire History of Australia.* Henry Holt and Co., New York.

Rose, D. (1994). *Country in Flames.* Proc. (1994) Symp. on Biodiversity and Fire in North Australia. Biodiversity Series Paper No. 3. North Australia Research Unit.

Tothill, J. C. (1993). Fire in black speargrass (*Heteropogon contortus*) pasture—a ten year comparison of burnt and unburnt treatments. In: *Proceedings of the Second Queensland Fire Research Workshop.* (Ed B. R. Roberts.) Darling Downs Inst. Adv. Educ. Press., Toowoomba.

Williams, R. J., Cook, G. D., Gill, A. M. and Moore, P. H. R. (1999). Fire regime, fire intensity and tree survival in a tropical savanna in northern Australia. *Australian Journal of Ecology* **24:** 50–59.

Burning early in the dry season, or where fuels are still partly moist, can be undertaken even in heavy fuels along creek lines and other fire-sensitive vegetation types.

6. Burning operations

by Tim McGuffog, Rodd Dyer and Brent Williams

Previous chapters have dealt with the nature and impact of fire regimes and how different regimes can be used to manage landscapes. This chapter deals with the 'nuts and bolts' of how to light and manage fires, and with the laws that affect fire management in north Australia.

Burning operations can be either strategic—as part of a management plan to deal with future fire threats ('prescribed burns') or tactical—suppression activities carried out in response to an immediate fire threat.

What is prescribed burning?

Prescribed burning is the deliberate ignition of vegetation and the subsequent control of the limits of the spread of the fire to achieve a desired management objective.

Prescribed fires can be applied in response to wildfire hazards, increases in native trees and shrubs, accumulation of rank ungrazed pasture or exotic weed invasions. Each management objective requires the implementation of an appropriate fire regime. This means manipulating the intensity (timing, season, fuel load and curing), extent and frequency

of fires over time. On pastoral land, special grazing management may need to be considered.

What is fire suppression?

The risk of uncontrolled wildfires can be reduced using early fire to reduce the hazard and by establishing a network of firebreaks ready for any outbreak. However, when fast-moving uncontrolled wildfires do occur, the key for efficient control is rapid response. Most fire control or suppression operations involve indirect methods, fighting the fires by back-burning from firebreaks in the path of the wildfires. Graders and other earth-moving equipment are used extensively to establish these forward control lines. They should be in place in advance.

Prescribed burns can be used for a number of reasons including the suppression of extensive, uncontrolled wildfires.

Managing fire behaviour

Fuel load, cover, moisture and weather conditions such as wind, temperature and humidity have the most influence on fire intensity. These factors have to be considered in the design and implementation of prescribed fires. Low-intensity fires (1000–2000 kW/m) are sufficient for removing rank grass and reducing wildfire hazards. High-intensity fires (>2000 kW/m) are required for reducing tree and shrubs.

What fuel load and cover is needed?

Fire intensity increases with higher fuel loads and cover (Figure 6.1). Dry fuel loads between 1500–2000 kg/ha (dry matter) and fuel cover of at least 40–50% are needed to support a continuous, moderate-intensity fire. Higher-intensity fires to control trees and shrubs need fuel loads of more than 3500 kg ha and cover levels of more than 60%.

Fires will spread even if fuel loads are low as long as fuel cover is continuous. If grassy fuel becomes more discontinuous, particularly from grazing, the rate of fire spread will decrease and fires will become patchier.

What influences fire behaviour?

Fuel loads and accumulation rates vary between pasture communities. Average seasonal conditions produce sufficient growth for fire to occur every 1–3 years in tropical woodlands and 2–5 years in semi-arid woodlands.

Accumulation of high, continuous fuel loads after above-average rainfall seasons allows prescribed burning with higher fire intensities or increases wildfire risk. Fuel is rarely available during drought years, particularly in heavily stocked grazing lands.

Increasing grazing pressure reduces fuel load, fuel continuity and fire intensity. Flame heights are lower in grazed pastures, reducing the impact of fire on the tree and shrub canopy. Dry fuel loads may rarely exceed 2000 kg/ha in savanna types with low and variable rainfall or when stocking rates are higher (Figure 6.2). In some cases, spelling for a growing season may be needed to accumulate sufficient fuel.

Figure 6.1 Fuel load and fire intensity

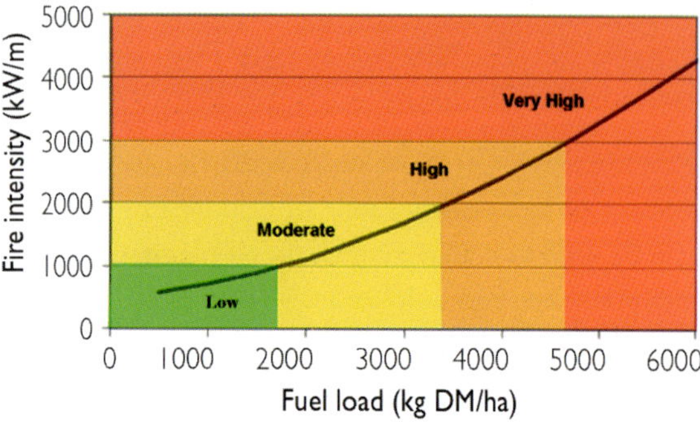

Increasing fuel loads increase fire intensity for grasslands. This relationship will be modified by factors such as wind speed, fuel moisture and fuel cover.

Figure 6.2 Stocking rates influence burning opportunities

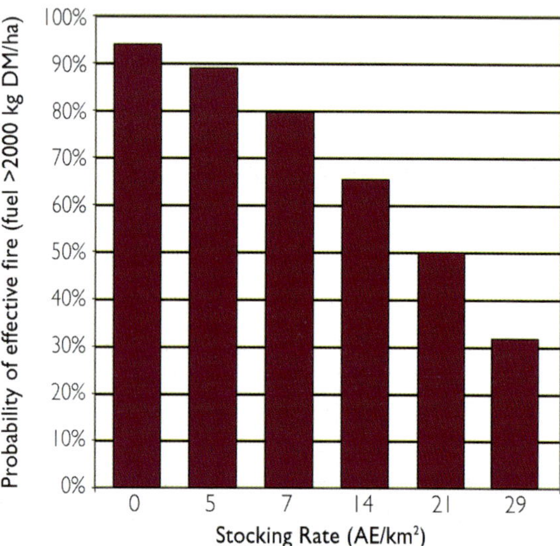

Increasing stocking rates decrease available fuel loads and reduce the chance of effective prescribed burns.

Is weather and the timing of burning important?

Wind speed, air temperatures, humidity levels, fuel curing and grazing significantly affect fire behaviour, particularly fire intensity.

Temperature and humidity

Air temperature and relative humidity values change greatly throughout the year (Figure 6.3). Lower-intensity fires generally occur when air temperatures are below 25°C and relative humidity is higher than 50%. Burning when air temperatures are above 30°C and relative humidity is lower than 30% will increase fire intensity.

Relative humidity remains low between July and October and increases rapidly from November onwards with the onset of the wet season; however, there is considerable variation between locations in northern Australia. This humidity reduces the potential of high-intensity fires for woody vegetation management. Although late dry season fires tend to be more intense than those in the early dry season, there is significant variation in 'fire weather' between years.

Wind speed and direction

Constant wind speeds of at least 10 km/h are best for continuous, moderate- to high-intensity fires and burning should not be carried out when wind speeds are greater than 20 km/h. Between September and October, prevailing south-easterly winds change to northerly winds that are often gusty and frequently change direction, making it difficult to predict fire behaviour.

Fuel curing or drying

The degree to which fuel is cured during the early dry season will vary considerably between pasture types. Due to their rapid maturity and curing, pastures dominated by annual and biennial grasses can generally support effective fires earlier in the dry season than perennial grass pastures.

Grazing effects

If grassy fuel remains relatively undisturbed by grazing, fire intensity will generally be greatest with late dry season burns. For annual pastures that cure early or are preferentially grazed early in the year, burning earlier in the dry season ensures more effective fires.

Figure 6.3 Temperature and humidity

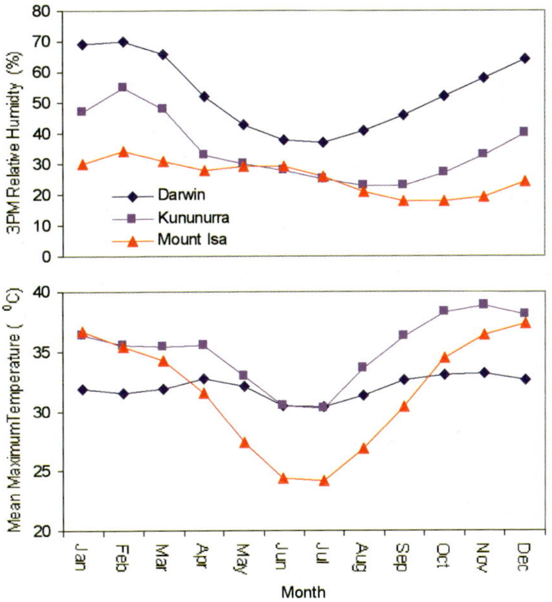

Monthly relative humidity and mean maximum temperature values change throughout the year and between locations.

High fuel load and cover after good seasons and with reasonable stocking rates can promote intense fires.

Grazing can severely reduce fuel load and cover, especially in poor seasons and on preferred pastures.

How does weather and fire behaviour change during the day and night?

Changes in temperature, humidity, wind and fuel moisture through the day and night (diurnal changes) greatly affect fire behaviour. Fires generally increase in intensity as the day progresses when temperatures increase and humidity drops. With the onset of dusk, temperatures drop, humidity increases and winds generally subside. This often causes fires to die down and possibly to go out, particularly during the late wet and early dry season when fuels have higher moisture content and heavy dew may form.

How should you manage grazing after burning?

In grazing lands, prolonged or heavy grazing soon after burning can damage palatable perennial grasses. Resting pastures from grazing after burning for a short period (weeks to months) or for longer (years) will allow grasses to drop seed and recover. However, it is often not practical to move stock in extensive lands. Rotational burning will shift grazing pressure throughout the paddock each year and minimise the risk of overgrazing previously burnt patches.

What are the risks to pastures from burning?

Besides the threat of a fire getting out of control, burning removes standing pasture that can be used for grazing. The risk of running short of pasture is reduced by burning late in the dry season or immediately after the first significant rains, and by checking that the area and frequency of burning match the average seasonal conditions, pasture type and grazing pressure. In regions where rainfall varies greatly between years, seasonal forecasting based on the Southern Oscillation Index may indicate the chance of a poor wet season and the need to preserve feed supplies.

NTDPIF

Fresh green regrowth can attract heavy grazing following burning. If burnt areas are not large enough, paddocks may need to be spelled to avoid overgrazing.

Fire Bans

Fire authorities declare Fire Bans in various ways across northern Australia. A Fire Ban means that no fires are to be lit in the open during the period of the declaration and in the area specified. The use of cooking fires is exempted under some legislation and other restrictions may prohibit the use of certain equipment or machinery. You should be familiar with the relevant state or territory legislation in regard to these restrictions.

Authorities may call a Fire Ban when weather conditions reach a point where the Grassland Fire Danger Rating is 'extreme'. The ban may be extended or cancelled as conditions develope.

Authorities may also declare a Fire Ban when they have a number of incidents in progress and resources are committed to the extent that they could not respond to further fires.

Permits to Burn

Each of the north Australian states has a 'Permit to Burn' system designed to coordinate and advise on the use of fires. Permits allow authorities to control the number and size of fires lit at any one time. They also advise local landholders and officials that the fires are authorised. There are established networks of local issuing officers who decide if conditions are suitable and identify controls and resources needed to manage the fire within specified boundaries.

Some agencies have cut-off periods for the availability of permits. Others allow the use of fires by permit throughout the year, but with increasingly tighter controls as the season dries out. In the broader pastoral areas, agencies may allow the unrestricted use of fires during the wet season and early dry season but require landholders to obtain a permit during a declared Fire Danger Period.

Contact your local fire authority and ensure that you are familiar with the local state or territory requirements.

Applying fire

The effective use of fire in landscape management is determined by the ability to cheaply, safely and easily ignite and control prescribed and suppression burns. The choice of ignition and containment options is critical in extensive areas as they affect fire behaviour, the impact on plants and animals and the ultimate success of fire as a management tool.

Prescribed and suppression burns can be undertaken from the ground or the air and can be applied with a number of ignition methods. Some of these are outlined below.

Ground operations

Prescribed and suppression burning operations on the ground can depend on many variables—the type of fuel, the stage of curing, topography, weather conditions and the type of burn desired.

The normal controlled burn is started in a downwind corner when burning in a paddock or block. This fire is then allowed to burn back into the wind before the upwind face is ignited. Wet season burns may be started at any point and allowed to spread naturally as the fire is self-extinguishing from moist fuel and ground.

Drip torches

The most widely used igniter is the 'fire bug' or 'drip torch' using kerosene. A mixture of one-third petrol, two-thirds diesel is often used but can explode. *Kerosene only is recommended*. The fuel dribbles over the dry grass and is ignited by the burning wick.

Home-made techniques include lighting the end of a piece of polythene pipe or an old tyre and dragging it along through the grass behind a vehicle. Lighting a fire this way can be dangerous if the vehicle becomes held up or breaks down.

'Wind/waterproof' matches—with an extended head which burns for long enough to be thrown into grass—are useful for point ignitions.

Capsule launchers

Capsule launchers can propel an incendiary up to 100 m from a vehicle into adjacent bush and have been used to widen firebreaks. Vehicle–mounted versions use compressed air for propulsion.

A cheap and effective launcher consists of a 1 m length of 40 mm PVC tube capped at one end. A primed capsule dropped into the tube and flung out by hand will travel a surprising distance.

Michael 'Magpie' Carter using a drip torch for back-burning operations

Flying in firebreaks at Carlton Hill Station, north-east Kimberley

To control wildfires on Carlton Hill Station just north of Kununurra, manager Geoff Warriner organises the Bush Fire Service (WA) plane to burn firebreaks in and around the property each year; he backs this up with a pattern of extra breaks put in from the ground.

Overall fire management plan

The key is aerial burning a firebreak around the perimeter of the property, with two other breaks burnt from north–south at intervals to divide up the property, and one from east–west across the middle. This creates six major blocks. Later in the season, additional breaks are burnt from roads and tracks to complement the aerial burning.

How do we put in the firebreaks?

Each year, we make arrangements with the Bush Fire Service for aerial incendiary burning of the boundary and internal firebreaks, and get our 'Permit to Burn' from the local shire.

We start burning as soon as the country will carry a fire, normally around Easter (March–April) and we stop in June.

The plane drops incendiaries to produce breaks which burn 100–1000 m wide before the fires self-extinguish (depending on fuel load and the amount of moisture remaining in the soil and pastures). The pilot is instructed to avoid the path flown for the previous year's breaks and any areas burnt during the previous season, either by planned or unplanned fires.

After the aircraft work is complete, we burn additional breaks from roads and tracks all over Carlton Hill until June when it normally becomes too dangerous to continue to light up. As with the aerial burning, wherever possible these breaks are put on country that was not burnt during the previous year, and the pattern is designed to achieve maximum effect in the prevention of wildfire.

What does this cost?

The main cost of burning in this way is the hire of the Bush Fire Service aircraft—currently $500 an hour. It normally takes three hours flying time to burn the breaks around and across Carlton Hill. The incendiary pellets are included in this figure, and there is also no charge for ferrying the aircraft and materials to the property.

The additional costs for ground-based burning are mainly for labour.

by Andrea Johnson

Fire ignited by incendiaries dropped from aircraft for strategic fuel reduction.

Flamethrowers

The flamethrower is usually a specially designed device mounted on a trailer or vehicle. A 12 V fuel pump pushes out a diesel-petrol mix from a 120 L fuel tank. Fire extinguishers and an emergency shut-off are provided for safety but the flamethrower should be operated only by trained personnel.

The flamethrower can be used to create hot fires for woody weed control or pasture development. It can also be used in wildfire control for lighting up over large distances in a short time as it can be operated at speeds up to 60 km/h.

Aerial operations

Aerial burning can be used to reduce fuel loads, create barriers to late-season wildfires and to ensure diversity of habitat. They are usually undertaken by bushfire agencies—on behalf of the landholder.

Aerial burning can be carried out in conjunction with ground operations to increase the efficiency of rotational burning programs on pasture country, particularly in extensive pastoral lands. The time spent implementing and monitoring prescribed fires until reasonable 'burn out' is achieved using only ground ignition methods in large areas can be significant. Aerial burning in a grid pattern following a perimeter ignition from the ground would significantly decrease the time for 'burn out', reducing the time spent at the fire.

Dropping incendiaries from aircraft

Incendiary devices can be dropped from aircraft to start a prescribed fire. A capsule or ball containing potassium permanganate is injected with a quantity of ethylene glycol and immediately ejected from the aircraft. Combustion occurs within 30–40 seconds allowing the capsule to reach the ground before ignition.

The incendiary devices and equipment have to be approved by the Civil Aviation Safety Authority (CASA).

Fixed-wing aircraft are used in the more extensive pastoral and conservation areas, being cheaper and faster, but helicopters allow greater precision and control on more subdivided land.

Helitorch

The use of Helitorch, a form of aerial ignition that delivers a continuous line of ignited gel, is effective for creating burnt firebreaks very early in the dry season and for controlling woody weeds such as *Mimosa pigra* on Top End flood plains. Its use is restricted as it is expensive and needs at least three trained staff to operate.

Aerial controlled burning operations

Aerial controlled burning operations have been used in the north since the early 1980s. They are generally limited to the tropical tall-grass areas although sometimes used to reduce fuel loads after a series of good seasons in the drier inland areas.

The bushfire agency works with the landholders toward the end of the wet season, agreeing on the location of drop lines and helping to develop a plan. Officers develop their regional and strategic plans in conjunction with adjoining regions. Each participating landholder must sign a contract and an indemnity that they are responsible for the control of the subsequent fire.

Aerial burning is normally undertaken from April through to June to produce 'burnt firebreaks'. These fires break up the landscape in recognised fire risk areas or reduce fuels over large areas of unused or unproductive country. The burns are usually timed to coincide with the initial curing of the grass at the beginning of the dry season, using green and wet areas to stop the fires. At this time, nights are becoming cool and there is still dew to extinguish fires.

Recently, aerial burning has been undertaken in the early wet season to encourage greater fire regime and habitat diversity on parks, reserves and defence lands.

Aerial ignition using Helitorch

Headfires burn with the wind and are more intense.

Headfires or backfires?

Many fires used in prescribed burning are lit as headfires. Headfires are lit upwind and burn with the wind direction. They are generally more intense and project heat energy upwards and, compared to backfires, impose greater damage to tree and shrub canopies. They also have higher rates of spread making burning in low and disturbed fuel loads more effective, while reducing the potential for damage to the pasture understorey.

Backfires are lit downwind and generally burn slowly into the wind. Backfires are used for lighting protective back-burns to save pasture on the downwind side from burning. They are especially important for controlling the progress of wildfires and ensuring safe containment of prescribed burns.

Point, line or perimeter ignition

Fire can be applied to an area in a number of ways (Figure 6.4). The choice of point, line or perimeter ignitions is important in terms of fire behaviour and vegetation response but the choice will be determined by the overall objective of the burn.

Point ignitions are lit from one or several points and allowed to spread out. They are generally less intense, patchy fires, often leaving islands of unburnt and partially burnt fuel.

Point ignitions are often recommended for conservation areas as fires spread out in all directions from a point allowing the fire to develop a range of intensities as it spreads. If fuel loads are discontinuous or areas for burning are very large, multiple point ignitions can be set in a grid pattern, either from the ground or air.

Line ignitions are applied with a continuous line of fire along one or several fire fronts. Line ignitions on several sides of a burn area become a perimeter burn. These fires are of higher intensity with higher rates of spread making them effective for tree and shrub management. Although they are relatively predictable and easy to manage, perimeter fires are less patchy and leave few refuges for wildlife and domestic animals to escape or resettle.

A fire regime where fire intensities are uniformly too high may cause undesirable effects on habitat, plant and animal diversity.

Firebreaks

The use of firebreaks, whether constructed or natural, enables fire managers to contain and control prescribed fires within an intended area. Several options exist for firebreak preparation, each with advantages and disadvantages.

Figure 6.4 Ignition methods

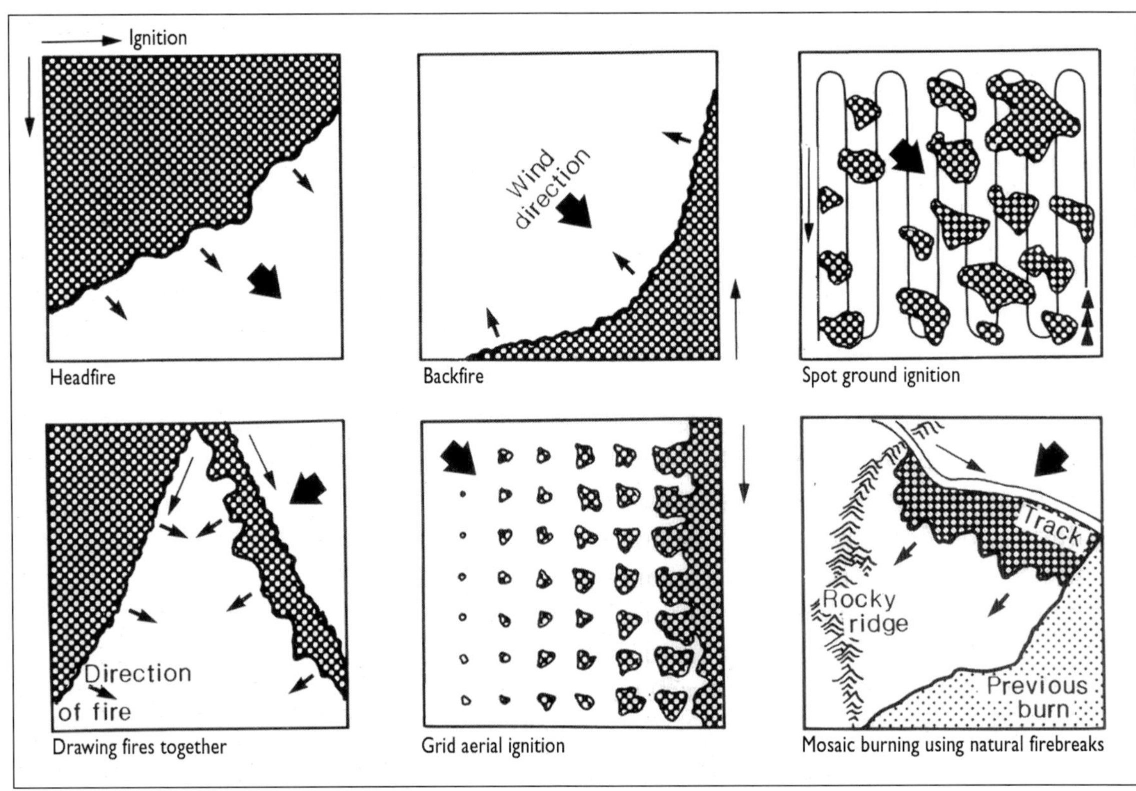

A variety of methods can be used to ignite and control fires during prescribed burning. Reproduced by permission of CSIRO

In practice, the main objective of firebreaks in extensive lands is not so much to stop the progress of oncoming fires, but to provide a line clear of flammable material from which to conduct protective back-burning or initiate management headfires.

Tactical or strategic firebreaks?

Tactical firebreaks are put in to control an existing fire. Strategic breaks are part of a total land management plan. Strategic firebreaks should take into account topography, soil type and erosion—existing boundary fence lines do not always do this.

Natural firebreaks

Natural firebreaks include streams, rivers and other wet areas where a fire will not burn and areas carrying little fuel such as rocky outcrops or recently burnt areas. Patch or mosaic burning strategies can use natural firebreaks such as creeks and ridges or previously burnt areas to control fire spread.

The effective control of fires relying on natural firebreaks can be improved by the careful timing of burning in relation to season, fuel curing state, time of day and prevailing atmospheric conditions. For example, fires lit during the late wet season (March–April) or late afternoon may extinguish at night due to falling temperatures and formation of dew.

Mechanical firebreaks

Mechanical firebreaks can be created using a grader, disc harrows, slashing, rolling or flattening. In heavy grass, the trash has to be removed to ensure that the protective line is not breached. Increasing fuel loads, fire intensity, presence of trees within 20 m of firebreaks, and decreasing firebreak width will increase the chance of firebreak breach by oncoming fires. The width of an effective firebreak will depend on these factors and could be as much as four or five grader blades wide in higher rainfall regions of the NT.

Ideally firebreaks follow the ridges to minimise soil erosion. When constructing firebreaks, the opportunity for run-off to channel should be minimised by spreading out the windrows and putting in diversion banks every 150 m across any firebreak running up and down any 1° slope.

Chemical firebreaks

Chemical firebreaks can be constructed along fence lines or around the perimeter of areas designated for burning. Herbicides, such as glyphosate, applied to growing pasture during the wet season create a dead, cured strip that can be burnt, leaving a clear break surrounded by a green, unsprayed area. Using herbicide can be a cheaper option than using machinery.

Benefits of using chemical firebreaks include:

- can go where grader or dozer disturbance would be detrimental
- wider firebreaks (up to 19 m swath width at a single pass) depending on equipment
- less soil disturbance
- generally cheaper
- can spray during the wet season and then burn off before adjacent areas are cured
- can manipulate fuel curing where other methods are impractical

Herbicide should be applied towards the end of pasture growth between late February and early March. This ensures that treated, grassy fuel is killed and 'dried off' while surrounding grass is still green and safe burning conditions exist. The burning of firebreaks is best done during late March or early April and in the late afternoon when weather conditions promote a 'hot' clean burn.

There are many types of sprayers. One of the most useful types is the boomless jet with high-pressure pump. This can be supplied by backpack tank or tank mounted on a trailer and towed by a quadbike.

Graders can put in mechanical firebreaks quickly.

Herbicide can be cheaper than machinery.

Fire management on Opium Creek, NT

Jeff Little manages Opium Creek Station some 180 km east of Darwin. The 200 sq. km property borders the Mary River system and the proposed Mary River National Park. Eucalypt woodland covers 60% of the station; the rest of the property is on the flood plain. As uncontrolled fire can have a devastating effect on the fragile, native flood plain pastures, fire management aims to reduce the threat of wildfires. Fire provides a cost-effective means of land management and weed control.

Fire management and the community

Fire management at Opium Creek is a major part of station management but also involves the community. As reducing the risk of wildfire is a community responsibility, Jeff is in the Point Stuart Volunteer Bushfire Brigade.

The results?

Jeff sees many benefits from his fire management strategy. They include:

- fewer wildfires

- protecting the fragile, native grass species on the flood plain

- less soil erosion on the chemically treated firebreaks

- better overall pasture management and property management

- reduced costs of replacing fences

- lower labour costs in fire management

The overall fire management plan

The key is the combined use of chemical application and fire to create firebreaks in strategic areas—these include fence lines on the dry scrub country and buffer strips bordering the flood plain between the high and low country. Three rocky ironstone outcrops on the property can create a real threat from wildfires due to lightning strike so they are control–burnt with buffer strips.

Practical methods on the property

Spraying buffer strips and fence lines

Herbicide is applied towards the end of the wet season while the vegetation is still lush and green. A single pass 10 m wide to either side of the fence is sprayed using a high-pressure boomless nozzle; this is mounted on a specially designed trailer/spray unit and pulled along at 12 km/h behind a quadbike.

Spraying the herbicide at the recommended rate speeds up the curing of the vegetation but leaves the root system of the grass and other species intact, thus lessening the risk of soil erosion.

Lighting the firebreaks

The sprayed grass is burnt when it has cured. The breaks are lit from the station quadbikes—with the local Point Stuart Volunteer Bushfire Brigade in attendance to help with fire control operations. The grass and other vegetation grows back after about 6–8 weeks, providing ground cover to avoid erosion but not enough growth to be a fire hazard.

by Jeff Little

Jeff Little, manager, Opium Creek Station

How to burn a paddock

1 Light simultaneously two backfires, burning against the wind, proceeding from the starting point to corners A and B of the paddock.

2 Allow these backfires to burn until an adequate firebreak has been established.

3 Light a headfire, burning with the wind, proceeding from points A and B to point C simultaneously and as swiftly as possible.

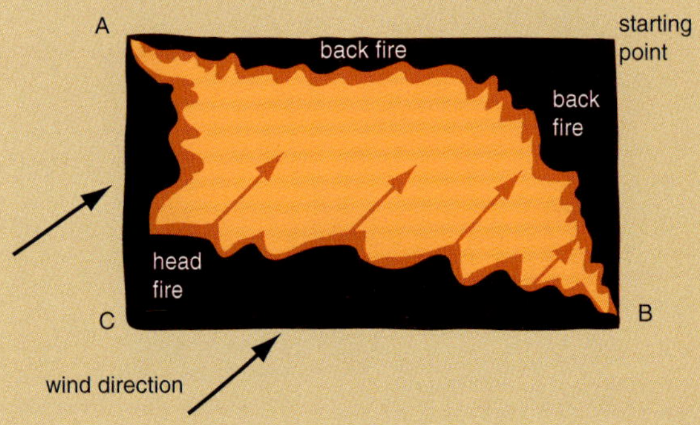

Prescribed burning on pastoral land

Rotational burning within paddocks is an effective way to use fire on pastoral land, particularly where there are multiple objectives for using fire. It involves establishing strategic firebreaks within paddocks, dividing them into sections that can be burnt in sequence over a number of years. The proportion, season, frequency of burning will depend on the specific objectives (Table 6.1).

Rotational burning provides a number of significant benefits to pastoral production. Regular fire will help maintain tree–shrub structure and density. Wildfire hazard is also reduced as fuel loads are reduced and fragmented, impeding the spread of uncontrolled fire through the landscape. Removal of old, ungrazed pasture provides fresh nutritious regrowth to cattle and may even-out grazing across the landscape. This will also reduce the disturbance of fuel loads in sections designated for burning during the next season. The invasion of exotic weeds is reduced because all areas of a paddock are burnt over a number of years which allows invading weed seedlings to be more easily spotted and removed.

Where to burn?

Burning should be initially directed to areas that are consistently ignored by cattle, where wildfire hazards exist, or where native or exotic woody weed problems are emerging. Burning should be avoided in pasture communities that are heavily grazed by cattle or that are in poor condition.

When to burn?

This depends on your objectives. Some general guidelines are outlined in the table below. Burning during the late dry season will promote high-intensity fires that will have maximum effect on removing woody plants. Early dry season fires will be less intense and more easily controlled and can be used to maintain the vegetation structure while allowing for fuel reduction. A combination of early and late dry season burning in paddocks during the one year can be utilised to serve several management objectives.

How much and how often to burn?

It is important to burn at least 15–20% of total paddock areas. If burnt areas are too small, cattle grazing pressure will be concentrated, potentially causing long-term damage to pastures due to over-grazing. Moderate stocking rates will allow controlled burning of a proportion of most paddocks in most years without affecting the supply of standing feed to cattle. A diagram showing rotational burning and other paddock burning strategies is shown on page 72.

A general guide to the size and frequency of paddock fires for different rainfall zones is shown in Table 6.2. In wetter northern areas (>800 mm) where burning frequencies would be between 3–5 years, paddocks could be divided into thirds or quarters and a section burnt each year. In semi-arid areas (400–800 mm) where desired fire frequency is lower, paddocks could be divided into quarters or fifths and 20–25% of paddocks burnt either each year or following above-average years.

Table 6.1 Recommended burning conditions for prescribed burning on pastoral lands

Management objective	Fire intensity	Fuel load (kg DM/ha)	Season of burn
Maintaining woody vegetation structure	Moderate–High	2000–3000	April–October
Change woody vegetation structure, control exotic weeds	High–Very High	2500–4500	August–October
Hazard reduction—reducing risk of wildfire	Low–Moderate	>1500–2000	April–June
Hazard reduction and provide early, dry green pick for grazing *	Low–Moderate	>1500	March–April
Remove old, rank pasture, modify grazing distribution	Low–Moderate	>1500	November–December

** Burning during the late wet season to extend the period of green pick into the early dry is generally not recommended except in lightly stocked, extensive paddocks in high-rainfall areas.*

Table 6.2 Recommended fire frequency and size for paddocks

Rainfall zone	Fire frequency (interval between fires)	Proportion of paddock burnt each year
High-rainfall areas >700 mm	2–5 years	25–50%
Medium-rainfall areas 400–700 mm	4–7 years	20–30%
Low-rainfall areas <400 mm	6–15 years	15–25%

Is planning necessary for prescribed fires?

Prescribed fires should be carried out when fuel and weather conditions are appropriate for each management objective rather than planning for a specific date. Prescribed fires often need to be lit at short notice so you need to be well prepared. Prior knowledge of correct fuel and weather conditions, and readiness for igniting and containing the fires is critical. It is important to have existing plans detailing what areas need to be burnt, what type of burn is required, what fuel and weather conditions are needed, and to have an ignition plan and firebreaks in place.

Be prepared with the right equipment when undertaking prescribed fires.

Figure 6.5 Fire management calendar

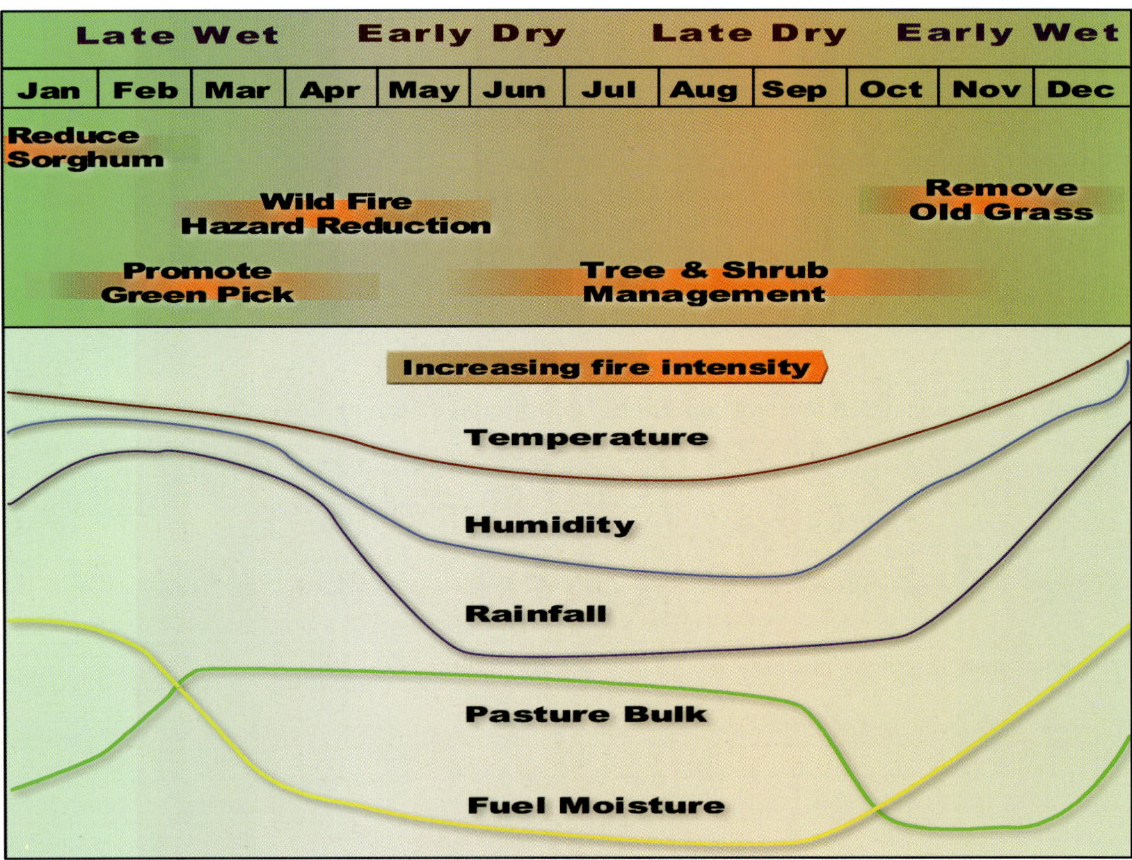

The seasons and months in which to burn for specific objectives and how this timing coincides with changes in temperature, humidity, rainfall, pasture bulk and fuel moisture

'Going burning' check list

Use this check list when preparing for a controlled, hazard reduction burn for grassland management.

1 Permit. Get a permit if necessary and abide by the instructions or advice.

2 Neighbours. Check with your neighbours. Do you know them or are they absentee? What is the minimum notice you need to give?

3 Values. What values on neighbouring land need to be protected?

4 Purpose. Why do you want to burn—fuel hazard reduction, development, shrub control?

5 Maps. Prepare and provide suitable maps for all sector bosses. Check that they are current and have a scale and legend.

6 Fuel type. What is the predominant vegetation—grass, scrub, trees? How is it likely to react?

7 Topography. How will topography affect the fire's rate of spread and access?

8 Area. What area is to be burned? How long will it take?

9 Weather. What is the general and local forecast—wind speed and direction, temperature and humidity? How well cured is the grassland?

10 Firebreaks. Know their width, position and direction, and mark them on the map. Are they accessible and cleared to bare earth? Where are the fall-back areas, are they natural clearings or man-made breaks?

11 Water points. Where are the refilling points, are they overhead, stand-pipe, dam or creek?

12 Manpower. How many workers are needed to manage the fire and for mopping up? Is there any backup or stand-by? How well trained are they? Will they be safe?

13 Safety. Demand everyone wears safety apparel at all times. Are first-aid kits are on hand and complete? Is there a trained first aider?

14 Victuals. Where is the drinking water and other sustenance?

15 Equipment. What equipment is required? Is it readily available in safe working order? Do workers know how to use it correctly?

16 Communications. Demand correct radio procedure and protocol. What are the capabilities of the radio equipment? State reporting requirements. Keep everyone informed.

17 Lighting pattern. Where do we start burning? Who will light up and when? Watch the security of downwind edges.

18 Mop up and patrol. Are additional manpower and resources required? Will the burn be completed and declared 'blackout' before the next day? Who is responsible to declare the burnt area 'safe'?

19 Wash up. Debrief personnel. Identify any problems that arose and correct them for the future. Publicly praise all for a good effort; privately upbraid those found wanting!

Suppressing fires

When suppressing fire in difficult terrain or in lightly stocked paddocks, back-burning will often be initiated from the nearest natural firebreak or existing cleared lines forward of the fire front, such as fence lines, roads and strategic firebreaks. As cleared breaks are established ahead of the fire front and flanks, back-burns are put in place around the graded perimeter effectively encircling the fire, removing the fuel from its path and stopping its progress. Although larger areas may be burnt out, this is often the most efficient and safest means of controlling large, established landscape fires.

Tactical firebreaks combined with back-burning may have to be rapidly installed in cases where existing firebreaks are far from the fire front, when the fire is still relatively small and controllable or where the value of unburnt land and pasture is high, and where firebreaks can be easily constructed. These firebreaks are thoroughly but rapidly constructed either ahead of the fire front or along both flanks of the heading fire, limiting its extent and rate of spread, reaching the fire front to finally extinguish its progress. This is called a parallel attack.

Be prepared

A rapid response to a wildfire requires having the appropriate equipment, ensuring that it is well serviced, functioning, ready for use and strategically located, and having personnel familiar with its use. Pastoral properties usually adapt station utilities with slip-on fire units or use water trailers whereas brigade vehicles are usually purpose-built 4WD vehicles or GFUs (grassfire fighting unit) carrying about 500 L of water. For larger fires, tanker trucks are used primarily as 'motherships' to fill smaller units.

Avoiding property damage from fire requires early season hazard reduction, planning and rapid response

A well-equipped firefighting unit with the Bushfires Council of the NT

Large tanker 'mothership' used to assist smaller firefighting units

Strategic fire management on Mataranka Station, NT

Over the past three years, Gary Riggs (manager), Kevin Wallace and Geoff Baker have developed a strategic fire management program on Mataranka Station about 90 km south of Katherine. The property is typical of the Mataranka and Sturt Plateau region with eucalypt woodland on red soils, but with a mixture of white sandy and gravelly soils.

Preventing wildfires

Firebreaks along the eastern boundary with its 56 km of frontage along the Stuart Highway consist of two breaks (each two blades wide) separated by 100 m. The middle section is burnt early in the dry season. About five weeks each year are spent on grading firebreaks, tracks and fence lines with another two weeks on the external firebreaks.

We burn a strip 200 m wide along the roadside during the mid to late wet season when the fire is self-extinguished by late afternoon humidity or storms. This is very effective and easy to implement and maintain as cattle grazing keeps the fuel load to a minimum throughout the year.

Other boundaries are sprayed with glyphosate early in the growing season (December) to drastically reduce the fuel load growth over the wet season. Spraying proves to be just as effective as grading, costs less and creates minimal soil disturbance. Glyphosate does not kill all plants, some of the vines keep growing.

Rotational burning

We first started rotational burning within paddocks in 1996 to reduce annual *Sorghum* and the density of young tree saplings as well as to reduce fire risk. Stock graze the freshly burnt areas making mustering much easier, while allowing the perennial grasses in the unburnt areas to build up root reserves and seed stores.

We burn one half of every paddock each year in the early wet season—after 100-150 mm of rain has fallen. This rain encourages the *Sorghum* to germinate so that it can be substantially knocked down while the fire is still hot enough to control the woody plants.

Fires are burnt fairly early in the day (usually 9–10 a.m.) to avoid storms in the afternoon and thus ensure a complete burn. We back-burn off tracks and fence lines using a quadbike with drip torch, then light up the whole perimeter section by section to avoid fire escaping into another paddock.

Costs of fire control

The costs of maintaining firebreaks on the station include wet season spraying, dry season grading and dry season roadside burning. These total $13,924 for the entire year—about $56 per km. Much of this cost is in grading major roads and fence lines which would be carried with or without a fire management strategy. The total cost for control burning is $2047 per year.

It is difficult to calculate the economical costs and potential losses due to uncontrolled wildfires. The relocation or agistment and supplementary feeding of cattle as a result of pasture loss is very expensive, while the loss of feed burnt could decrease productivity due to poor animal condition and lower calving rates. The benefits of fire control and rotational burning far outweigh the costs.

by Gary Riggs and Andrea Johnson

Gary Riggs, manager of Mataranka Station

Fire is used to control acacias on disturbed areas.

If I see a bushfire?

When people first see bushfires in northern Australia it can come as something of a shock—fires and smoke are frequent, fire trucks are rare and the country is often burnt and black. But fires are a natural part of the savanna landscape in northern Australia. Travellers still raise many questions. What do I do if a fire comes close? Should I report fires to the authorities? What can I do to help?

Landholders are responsible for fire management on their land, but many properties do not have the resources to conduct hazard reduction burns nor to control wildfires. They have to rely on their neighbours or the immediate community for assistance.

State and territory bushfire agencies provide varying levels of assistance to landholders for hazard reduction, coordination of fire management and firefighting when required.

Am I at risk from bushfires?

In most cases, no—the fires you see will not harm you if you know what to do and your property is well prepared. These fires are much less intense than the bushfires of southern Australia because they simply do not have the fuel to burn.

This is because much of the fuel (grass and tree litter) either decomposes or gets burnt each year. The amount of grass and litter in the north never rises above the equivalent of a few years' growth, in contrast to southern forests where litter can accumulate over decades.

The intensity of bushfires also depends on the time of year. There are few fires in the wet season with its heavy rains. Fires become common in the early dry season (May to July) but cooler conditions and other factors limit their intensity.

Outbreaks can be more dangerous, however, in the late dry season (August to November), when the grass and tree litter is drier, it is hotter and fires can be fanned by strong winds. These fires can be a threat if adequate precautions are not taken.

What if I encounter a bushfire while driving?

If the fire is just burning the undergrowth with little smoke you should not be in any danger. In these northern regions, even higher-intensity fires rarely put out enough heat to make roadways impassable.

If there is a lot of smoke...

- Turn on car headlights

- Slow down and be aware that there could be people, vehicles, large trucks and livestock on the road.

- Follow directions of police and firefighters if present.

- If you cannot see clearly, pull over to the side of the road, stop your vehicle, keep the headlights on and wait until the smoke clears.

Should bushfires be reported?

Fires in the early dry season generally do not need to be reported.

Many of these fires are lit to manage the land. To prevent destructive wildfires the grass and litter load is reduced before the hotter time of the year by lighting low-intensity fires in the early dry season in key areas.

Later, hotter bushfires, seen from August to November, can be more harmful to property and the environment and have greater potential to harm people. These fires should be reported to the appropriate authority (see Table 6.3).

This brochure provides travellers with information on bushfires

Volunteers

In all rural districts, successful fire management would be impossible without the contribution of volunteer firefighters. These men and women donate countless hours to the cause of fire management in their communities, not only fighting wildfires but, more importantly in tropical savannas, helping landholders implement hazard reduction programs which reduce the severity and impact of late dry season fires.

The safety of life and property and, in many cases, the health of local environments in rural areas relies to a large extent on the hard work of these volunteers.

In the tropical savannas, the role of volunteers in fire management is similar in Western Australia, Queensland and the Northern Territory although there are differences in the government support systems, funding processes and coordination. For example, in the Northern Territory volunteer bushfire brigades are independent incorporated bodies, but they receive operational funding support and are provided with firefighting units and tankers from the Bushfires Council through government grants. In contrast, Western Australian bushfire brigades are supported by a combination of state and local government resources.

Volunteer bushfire brigades often provide a focal point for social events within the community, and help to engender a sense of community spirit in rural areas. In the Northern Territory, brigades compete in and help organise an annual field day competition which brings teams from across the Territory together for an intense day of competition and socialising.

Brigades also often provide their services to help support other community groups and organisations, and form a valuable group of equipped and experienced personnel during times of civil emergency such as cyclones and floods.

A role for everyone

The primary function of volunteer bushfire brigades is to prevent and fight wildfires. There is always a demand for more volunteers who are willing and able to undertake tasks such as fuel hazard reduction, attending bushfires and operating firefighting equipment.

Public awareness signs help get the message across to the community.

Volunteer firefighters and landholders muster before tackling a large fire.

The start of the relay race at the NT annual Volunteers Field Day

Volunteers cool down during a break at a fire.

However, like any operational organisation, there are many tasks other than front-line fire management that are critical to the effective functioning of the brigade. These include equipment maintenance and repairs, fund-raising, organising promotional and social events, seeking sponsorship and support from local businesses and administration.

Volunteer bushfire brigades also rely on the active involvement and cooperation of landholders to undertake fire prevention works on their properties, to be prepared for wildfires and to coordinate these activities with their neighbours. A well-managed property with maintained firebreaks and reduced fuel zones around infrastructure greatly assists brigades when there is a wildfire to be fought and contained.

What training is available?

Volunteer firefighters are offered training appropriate to the tasks that they will be expected to undertake as members of a volunteer bushfire brigade. The skills and knowledge gained during training include fire behaviour, personal safety, fire suppression, how to operate tools and equipment, an understanding of the relevant laws and other issues useful to their activities. Advanced training is available to those who take on more senior and responsible positions within the brigade.

The training of firefighters has taken on national importance with the development of the Australian Fire Competencies. Training is now competency based; the curriculum and qualifications are nationally accredited and they can be articulated between the states and territories.

This training is promoted not only to the volunteer brigades and other land managers but also to rural people as a whole. Bushfire knowledge and skills is considered a life skill as important as first-aid training or learning to swim. Many volunteer bushfire brigades now require all their members to have this minimal qualification before they can go on the fire ground.

How do I join a volunteer brigade?

Contact your local bushfire management agency (Table 6.3) which will put you in contact with your local volunteer brigade.

Field Day races test volunteers' skills with the 'flaming bag' and provide friendly competition.

Volunteers test their equipment skills at competitions before the start of the fire season.

Bushfire training course.

Legislation

Legislation concerning fires in the landscape in northern Australia is administered by a number of different agencies. The main legislation regarding rural fire management is administered and enforced by relevant bushfire agencies in each jurisdiction. These are the Queensland Rural Fire Service, Bushfires Council of the NT, and the WA Fire and Emergency Services Authority respectively.

There is also legislation relating to state and territory land management agencies such as parks and wildlife agencies which controls the use of fire on their areas of management. Urban fire authorities also have responsibility for bushfire management around urban areas and have legal requirements for the use and control of bushfires; these can impose additional requirements on the landholder.

It is therefore important that you know within which jurisdiction your land falls, and be aware of the specific legislative requirements.

Relevance of fire legislation

In each of these states and the Territory, the general theme of fire management is that 'fire is the landholder's responsibility'. All agencies across northern Australia focus primarily on fire prevention measures as opposed to fire response and suppression.

Legislation has been drafted on this basis to enable landholders to use fire to manage their land responsibly. In each case, legislation puts in place a number of controls and restrictions to ensure that fire management is undertaken in a coordinated and sensible way, and to minimise the risk of fires escaping and causing damage to life and property.

These include the requirement to get permits to burn at various places and times, conditions and restrictions placed on the users in undertaking burning, and the requirement to advise neighbours and authorities where and when burning is being undertaken. Authorities may also impose fire bans during times of extreme weather conditions in order to prevent fires getting out of hand or to control the use of fire when resources are committed elsewhere.

Legislation also allows the agencies to enforce and undertake both fuel hazard reduction and to enforce establishment and maintenance of firebreaks. Some agencies also have provision for recovering costs associated with fire control operations.

The use of intense fires for woody plant control is becoming recognised as an essential land management tool. Agencies have various means to allow these fires to proceed during periods of high fire danger late in the dry season; however, there is still an onus on landholders to undertake early preparatory work in establishing firebreaks and ensuring that these fires remain under control.

Urban-rural interface

Agencies managing fires on the urban interfaces around the various communities and town centres in northern Australia usually impose tight restrictions on the use of fire for land management purposes. This is because there is usually a high risk of damage to property and assets in these areas and there is a far greater input required to coordinate smaller landholdings.

Volunteer bushfire brigades are established by all three north Australian agencies. However, each is established and administered differently and has different levels of local and state government involvement. Most volunteer brigades are formed in the more populous rural interface areas, though brigades also exist in the broader pastoral areas. Volunteer bushfire brigades are an invaluable community resource which allow agencies to better coordinate fire prevention and control programs using local skills and knowledge. Landholders can combine their resources and help each other to undertake their fire management programs.

Your local fire legislation and authority

Each state and territory is listed in this document (see Table 6.3). If you are unaware, or need further information about fire management, you can contact the relevant agency which will give details on local contacts for permits, fire management advice and contact details for volunteer brigades in your area.

Each state and territory has differing mechanisms to implement fire restriction or prohibition times and areas and you must be aware of these regulatory controls before lighting any fire.

Unplanned fires occur regularly and it is in your best interest to know whom to contact quickly in the event of a bushfire affecting your property.

Table 6.3 Information on state and territory fire agencies and legislation

Principal State/Territory Authority	Bushfires Council of the Northern Territory
Managing Agency	Bushfires Council of the NT
Legislative Basis	*NT Bushfires Act*
Principal North Australian Office	1718 Albatross St Winnellie NT Ph (08) 8922 0844
Principal Officer	Chief Fire Control Officer
Web Address	http://www.nt.gov.au/bfc/
Regional Agency Officers	Regional Fire Control Officers
Voluntary Regional Authorities	Fire Wardens
Over-riding principle of all agencies	Fire suppression is the responsibility of landowner or occupier in each
Permits to burn	Required in a fire protection zone and during a fire danger period in a declared fire danger zone
Conditions of permit	Vary significantly between the states and territories and certain local
Fires for which permits are not required Note: Strict and various conditions for lighting these fires exist	Cooking or camp fires, disposal of animal carcass
Permits issued by	Fire Control Officer or Fire Wardens
Volunteer brigades established, equipped and maintained by	Bushfires Council of the NT
Assistance provided to landholders	Equipment subsidy, aerial burning, fire prevention and control, training, communications
Regulations on the use of certain machinery and equipment	Controls on the use of oxywelding and cutting equipment, and requirements for spark arrestors and fire-extinguishers on machinery in the open
Fire restrictions and prohibition	Fire ban may be declared during very high or extreme fire weather conditions for specified area and time

Further reading

Cheney, N. P. (1981). Fire Behaviour. In *Fire and the Australia Biota*. (Eds. A. M. Gill, R. H. Groves and I. R. Noble.) Australian Academy of Science, Canberra. pp. 151–175.

Cheney, N. P., Gould, J. S. and Catchpole, W. R. (1993). The influence of fuel, weather and fire shape variables on fire-spread in grasslands. *International Journal of Wildland Fire* **3(1):** 31–44.

Cheney, P. and Sullivan, A. (1997). *Grassfires: Fuel, weather and fire behaviour.* CSIRO, Australia.

Gill, A. M., Moore, P. H. R. and Williams, R. J. (1996). Fire weather in the wet-dry tropics of the World Heritage Kakadu National Park, Australia. *Australian Journal of Ecology* **21:** 302–308.

Hodgkinson, K. C., Harrington, G. N., Griffin, G. F., Nobles, J. C. and Young, M. D. (1984). Management of Vegetation with Fire. In *Management of Australia's Rangelands.* (Eds G. N. Harrington, A. D. Wilson and M.D. Young. CSIRO Publications. pp. 141–156.

Fire and Emergency Services Authority of Western Australia	Queensland Fire and Rescue Service
Bush Fire Service of WA	Qld Rural Fire Service
Bush Fires Act WA, FESA Act	*Qld Fire and Rescue Authority Act*
Unit 4, 16 Headland Pl Karratha WA Ph (08) 9143 1227	31 Grove St Cairns Qld Ph (07) 4039 8240
Executive Director Fire Services	Rural Fire Service Commissioner
http://www.fire.wa.gov.au/	http://www.ruralfire.qld.gov.au
District and Operations Managers	District and Regional Inspectors
Local Government, Bushfire Control Officers	Chief Fire Warden, Fire Wardens
state and territory. Agency's main focus in all areas is on fire prevention rather than fire suppression	
Permits required during restricted burning times declared by shire and town councils	Permit or notification to burn required for all fires lit in the open
conditions may be required by issuing officers	
Cooking or camp fires, disposal of animal carcass, garden refuse, sawmill residue or use of incinerator	Enclosed cooking fires, disposal of animal carcass, disposal of sawmill residue involving material less than 2 m in height and diameter. Special exemptions for sugar cane Industry
Bushfire Control Officers	District Inspectors or Fire Warden
Local Governments	Qld Rural Fire Service
Equipment subsidy, aerial burning, fire prevention and control, training, communications	Equipment subsidy, aerial burning, fire prevention and control, training, communications
Controls on the use of oxywelding and cutting equipment, and requirements for spark arrestors and fire-extinguishers on machinery in the open	Controls on the use of oxywelding and cutting equipment, and requirements for spark arrestors and fire-extinguishers on machinery in the open
Prohibited burning times declared for specific zones and periods each year. May include vehicle and harvesting restrictions	Local fire ban imposed on entire or part local government area for a specified period

Luke, R. H. and McAthur, A. G. (1978). *Bushfires in Australia*. Aust. Govt. Publ. Serv., Canberra.

Williams, R. J., Gill, A. M. and Moore, P. H. R. (1998). Seasonal changes in fire behaviour in a tropical savanna in northern Australia. *International Journal of Wildland Fire* **8(4):** 227–239.

A working knowledge of fires and their impacts is just one part of effective fire management. However, you also need to be able to adapt your fire management in the light of experience—'adaptive management'. To do this you need to be able to monitor fires effectively.

This chapter describes tools which land managers can use for monitoring the effects of their fire management; it also describes the use of satellite-based technologies for obtaining information on the regional distribution and extent of fires, and on the state of curing of fuel. It examines how models can be used to provide useful information about seasonal and fuel conditions. A third section describes a range of map-based products that will assist managers in understanding seasons, fuel loads and curing state and fire risk at a regional level.

On-ground monitoring

While many experienced land managers develop a detailed understanding of the effects of their fire management, memories can be short-lived; there is often a high turnover of staff on cattle properties or conservation reserves. Useful monitoring observations, both positive and negative, need to be recorded and made accessible if they are to have lasting benefit. On-ground sites may be used to monitor the responses over time of key indicators such as soil cover, species composition and status, vegetation condition and woody thickening. Aerial photography and even historical photos have been used to assess long-term changes particularly in woody plant cover and density.

On-ground monitoring systems

The design of ground-based fire monitoring systems range in complexity from simple photo-points to a series of sites in a sophisticated, replicated experimental design. The design depends on the purpose of the monitoring. Photo-points at fixed sites have been used to record the occurrence of fire or to show vegetation and landscape changes over time. Simple plots may be established to record the immediate effects of a fire on the responses of individual species. These plots would include a

photo-point plus counts of the individuals of a single species, such as the weed rubber vine. The more complex monitoring sites are designed for long-term monitoring of regional trends in plant and animal community structure and composition. The data collected include information on the species mix plus biomass samples, and it requires considerable statistical skill to adequately analyse the results.

The use of photo-points should be done systematically by taking photos from the same spot and aimed at a marked reference point or target.

The plot-based fire monitoring programs, as established by various agencies across northern Australia, are resource hungry and require substantial ongoing commitment. A comparison of some long-term monitoring approaches used in Queensland, the Northern Territory and Western Australia is shown in Table 7.1.

Photos to monitor local change should be taken using a fixed reference point.

Table 7.1 Examples of formal, plot-based fire monitoring programs adopted by agencies in Queensland, Northern Territory and Western Australia

Attribute	Queensland	Northern Territory	Western Australia
Implementing agencies	Queensland National Parks & Wildlife Service	• Parks & Wildlife Commission • Parks Australia: Kakadu NP	Agriculture Western Australia
Purpose of monitoring scheme	To assess changes in vegetation health on national parks and effectiveness of off-park conservation agreements	To inform park managers of the long-term implications of imposed fire regimes on vegetation and fauna	To assess the effects of fire regimes on major Kimberley pasture types and inform the regional pastoral community
Year started	1998	1994	1993
Current state of implementation	• Over 100 plots established	• Over 250 plots in four Top End major national parks	• 50 plots which have been assessed at least twice
Monitoring rationale	Mostly issue-focused, favouring an experimental approach • Assessing appropriate fire regimes for control of woody plants in grasslands • Conservation of certain fauna	Mostly passive monitoring • Assessing trends of habitat and vegetation response to current fire regimes • Focus on a representative range of vegetation types, including critical fire-sensitive habitats	Mostly opportunistic monitoring • Comparing the vegetation response of intensely burnt pasture on opposite sides of a track • Documenting vegetation change in frequently burnt savanna woodland
Plot dimensions	50 × 4 m	40 × 20 m	50 × 26 m
Woody vegetation sampling	• All woody individuals identified and mapped as distance along and away from central axis; height and diameter at base measured	• Trees (>5 cm at 1.3 m height) in 40 × 20 m plot: all stems tagged, measured for height and diameter at 1.3 m, assessed for fire scars • Shrubs (>50 cm height and <5 cm diameter at 1.3 m) in central 40 × 10 m plot: counts per species in 2 height classes. • Small shrubs (<50 cm height) in 80 1 m² quadrats within plot: counts per species	• All woody individuals identified to species within 100 0.5 m² (70 cm × 70 cm) quadrats • Heights of juveniles recorded
Ground cover/ herbaceous species sampling	• Occurrence of target or all species in 20 systematically placed 25 cm × 25 cm quadrats	• Foliage cover (%) of each species in 40 systematically placed 1 m² quadrats	• Presence of all species within 100 0.5 m² quadrats • Composition estimated by dry weight, rank method
Other attributes monitored	• Ground-layer fuel using drop disc method at 50 points • Animal disturbance/signs recorded	• Fauna sampling undertaken on sites adjacent to vegetation plots	• Ground-layer fuel and moisture content from 15 0.5 m² harvested quadrats • Grazing and dung recorded in each quadrat
Time taken to inventory each plot in the field	• 1–3 hours for initial site establishment, shorter for re-measurements	• 2–3 hours for full inventory, plus plant specimen identifications and database	• 4–5 hours, including fuel sampling
Assessment frequency	• Annually or as frequently as sampling design requires	• Biennially for fire history • 5 years for vegetation	• Annually to biennially for fire history
Management issues	• Supported by field staff and stakeholder partners • Need a flexible system for looking at different land management issues • Still under development and needs long-term institutional backing	• Strongly supported by field staff • Good commitment from participating agencies • Very substantial personnel and logistical resource requirements	• Importance of maintaining communication with co-operating pastoralists • Matching site network to available resources; logistical constraints may prevent data collection before fires • Plant species identification

Monitoring pasture composition under frequent fires

Intermittently and lightly grazed savanna woodland near Kununurra, WA was monitored after a relatively intense fire late in the dry season of 1993. There were four subsequent, mostly unplanned, fires (1995, 1997, 1998, 1999) but all were of lower intensity.

Monitoring has revealed an interesting trend in pasture composition. Despite continuing disturbance by fire, the proportion of perennial sorghum (*Sorghum plumosum*) has been increasing relative to other less palatable perennial grasses. This may just represent a recovery from the effect of the hot 1993 fire but is considered beneficial from a pastoral perspective.

by Andrew Craig

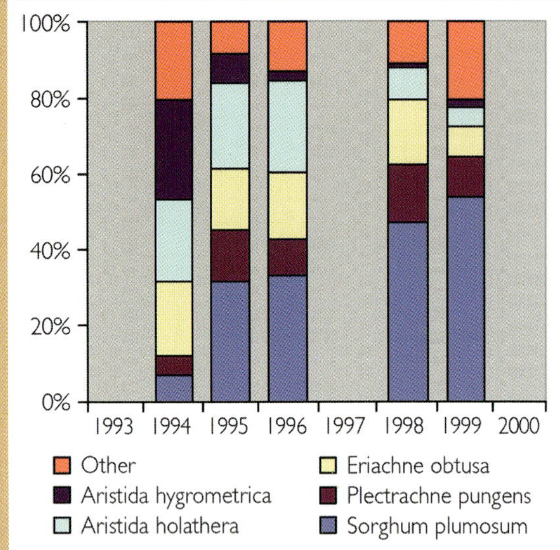

Changes in botanical composition of lightly grazed native pasture under regime of frequent fires

Monitoring soil loss after wet season burning

Annual sorghum grasses are a major source of dry season fuel in the Top End and Kimberley. Burning the annual sorghum in the growing season before the plants have set seed can reduce fuel loads and can allow other perennial grasses to increase. While wet season burning reduces sorghum, it does not appear to impact on other annuals. In fact, removal of sorghum allows annuals with soil seed banks to proliferate. Wet season burning is an increasingly applied approach for reducing annual sorghum, but the effects of soil loss also need to be considered.

Does it increase soil erosion?

An area of annual sorghum in the Department of Defence Bradshaw Field Training Area in the Northern Territory was burned in mid-December 1999 when the grass was growing but not seeding.

On each side of a valley, three plots were measured on the steep mid-slopes and three on the lower levees. One side of the valley was burnt with the other side left unburnt as a control. Vegetation was recorded for each plot before burning.

Erosion in each plot was assessed by hammering roofing nails into the ground along the contour at 20 cm intervals and measuring the amount of soil loss from under the cap of each nail at the end of the wet season.

Soil loss from burnt steep slopes averaged 6.7 mm, but losses were much lower on the burnt lower slopes and in unburnt plots.

Relatively high rates of loss from burnt steep slopes might be attributable to very heavy rain associated with Cyclone Max; however, care needs to be taken when burning steep slopes in the wet season.

by Cameron Yates

Further reading

Stocker, G. C. and Sturt, J. D. (1966). *Australian Journal of Experimental Agriculture and Animal Husbandry* **6**: 277–279.

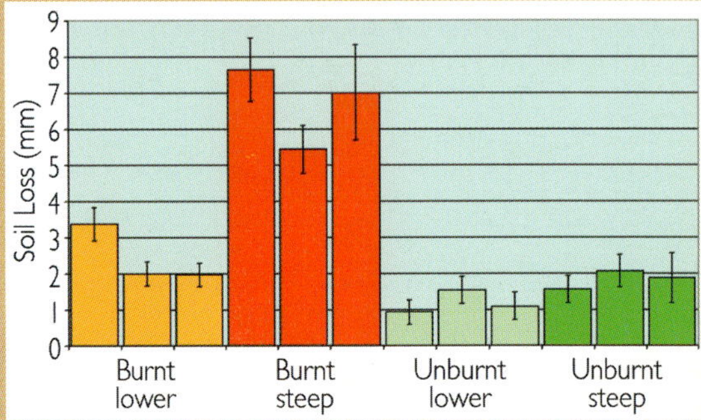

Mean soil loss (mm) from upper and lower slopes after burning annual sorghum during the wet season. Three replicate plots were monitored for each burning treatment.

Aerial transects flown over western Arnhem Land for counting living and dead cypress

Counting rings and measuring stem diameter at 1.3 m height

Monitoring cypress pine stands in Arnhem Land

Cypress pine (*Callitris intratropica*) is a long-lived tree species that occurs across northern Australia in a range of habitats with free-draining soils.

High-intensity fires will kill or scar mature cypress trees but the termite-resistant stems remain standing for many years. Low-intensity fires may kill juvenile trees but will not generally affect mature trees; thus the condition of cypress stands can indicate the severity of the fire regime.

In recent years, ground-based cypress surveys over many parts of northern Australia have counted living and dead pine trees to determine their current condition and former range. Both living and dead cypress are highly visible in the savanna landscape, and aerial surveys have been undertaken to extend our knowledge base.

An aerial survey in October 2000 showed that the condition of cypress pine stands in western Arnhem Land varied considerably.

There were more living trees than dead stems in areas protected from frequent fire—in the rocky areas in the dissected sandstone and the escarpment country.

On the open sand sheets and savanna lowland woodlands, the dead stems far outnumbered the living, indicating that fire severity has increased in recent times.

The diameters of dead trees are measured, sections cut and the growth rings counted to determine a relationship between tree diameter and age.

by Andrew Edwards

Further reading

Price, O. and Bowman, D. M. J. S. (1994). *Journal of Biogeography* **21:** 573–580.

Hammer, G. L. (1981). *Australian Forestry* **44:** 35–41.

Living and dead cypress pines

Monitoring vegetation change

Rainforest in and around Iron Range National Park in far north Queensland is expanding into the grasslands—a result of restricted burning over the last half century.

Aerial photography's fine resolution makes it useful for monitoring this long-term vegetation change; photographs for the Iron Range National Park area are available for 1943, 1970 and 1991.

Vegetation communities are delineated on the photographs, and the line information is digitised for mapping and measuring areas.

In a 140 sq. km study site in Iron Range National Park, the rainforest increased by 33.2 sq. km between 1943 and 1970, and by a further 4.4 sq. km from 1970 to 1991.

These vegetation maps are being used in planning strategies for the management of the park and surrounding areas.

by Peter Stanton and Lisa Roeger

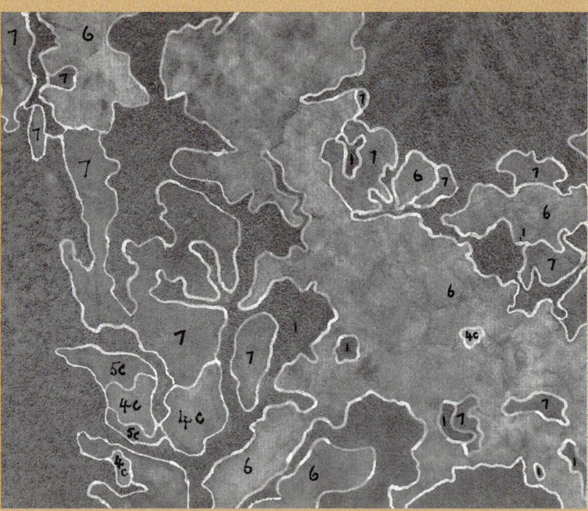

Examples of aerial photographs of Iron Range National Park for 1970 (above) and 1991 (below). The rainforest can be seen as the darker areas.

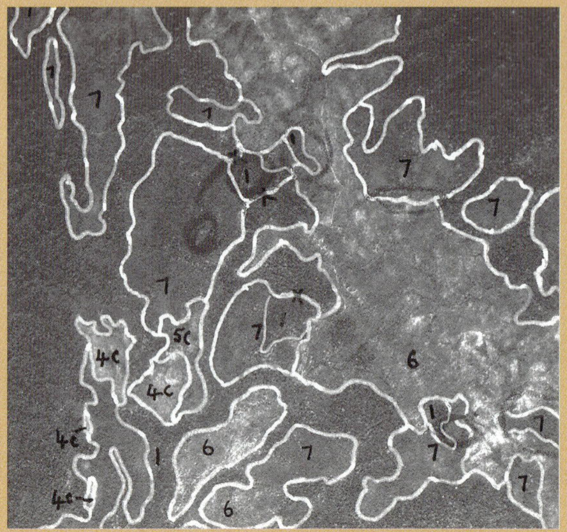

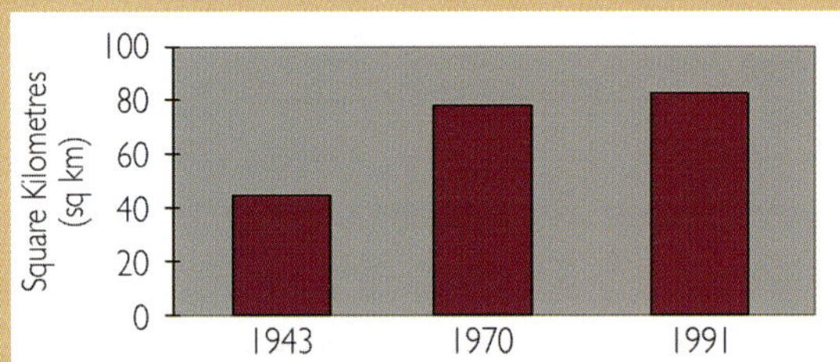

Changes in the area of rainforest in Iron Range National Park between 1943 and 1991

Overlaid digitised boundaries of rainforest in Iron Range National Park showing the expansion of the forest area. The original area in 1943 (dark green) has expanded to the larger area (lighter green) in 1970 and 1991.

Aerial photo monitoring

Aerial photo records are available for many parts of northern Australia from the 1940s, the 1960s and often more recently, as part of national or regional mapping projects. While aerial photos are typically taken too intermittently to be useful for describing fire regimes in savanna systems (although effective for mapping fires in more arid regions) they can establish long-term trends in woody vegetation thickening or clearing (see p. 106).

Figure 7.1 Willowra area Landsat image

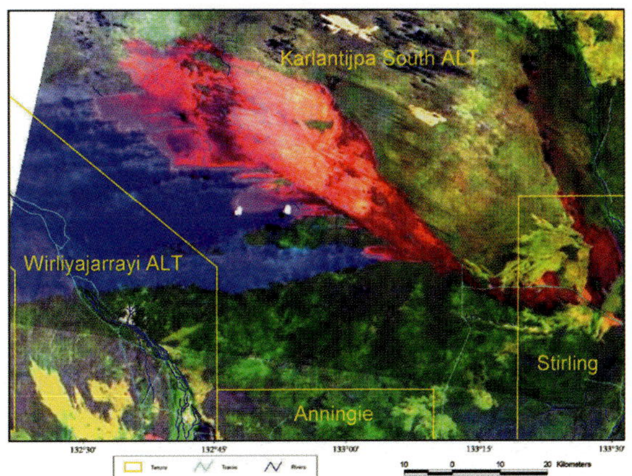

The fire was started about 28 March 2001, six days before the Landsat image on 5 April 2001. The recently burnt areas are red. Previously burnt areas, from fires in October 2000, are the yellow patterns in the north-east and south-west parts of the image. The two small whitish patches in the central north portion of the image were fires between 30 December 2000 and 16 February 2001.

The fires were probably started by lightning but subsequently put out by associated rainfall before large areas were burnt. The small dark patterns north of the fire are remnant mulga communities within the extensive spinifex sandplain area that dominates the rest of the image. The Lander River flows north through the Wirliyajarrayi Aboriginal Land Trust in the south-west portion of the image. The Willowra community is just off the southern margin of the map on the Lander River.

The fire was lit from a track just west of Stirling Station. For the first few days the fire burnt slowly before south-east winds pushed the fire to the north-east. A wind change from south-east to east on the day of the satellite pass turned the western flank into the fire front. The blue haze to the west of the burnt area is the smoke plume. The pink edges to the fire are the burning fire fronts and the two white dots are clouds forming on the smoke plumes above the fire front.

The area burnt at the time of the image was 1500 km².

The fire continued to burn for the next three months. The final extent of the fire was approximately 10,000 km².

Satellite-based monitoring

At broader regional scales, fire regimes and their ecological responses can be monitored using satellite technology. Satellite images combined with computer mapping software (or Geographic Information Systems—GIS) allow the trends and responses observed from ground-based plots to be extrapolated to the scales of properties, land systems, regions and even continents. The use of satellite data has become an important part of monitoring both fire regimes and vegetation response to fire in the tropical savannas. Satellite images are used to detect active fires, map the extent of burnt areas and assess the greenness or curing state of vegetation. This section has technical information on the two main satellite systems currently used in fire monitoring. It is provided as background information to help users understand the issues associated with the use of satellite data for fire and vegetation monitoring.

Satellite systems

Two satellite systems are currently used for fire monitoring and mapping—NOAA (National and Oceanographic Atmospheric Administration) and Landsat.

NOAA

Originally designed for mapping weather patterns and cloud formations and measuring sea surface temperature and patterns of ocean circulation, NOAA's primary sensor, the AVHRR (Advanced Very High Resolution Radiometer), also provides useful land surface information. The four currently operational satellites (NOAA-12, -14, -15 and -16) continuously orbit the Earth and cross part of Australia twice a day.

Landsat

Landsat 7 was designed specifically for mapping and monitoring the Earth's resources using the ETM (Enhanced Thematic Mapper) sensor. The sensor has three components—panchromatic, multispectral and thermal—with different spectral and spatial capabilities. An example of a large fire scar detected by Landsat is shown in Figure 7.1.

Satellite orbits

Both NOAA and Landsat satellites have similar orbital characteristics—the main difference being the time of day when the satellite collects data. Landsat 7 collects data only on the sunlit side of the Earth and crosses the equator at 1000 local solar time. The NOAA satellites collect data on both sides of the Earth (daytime and night-time passes). Information from sunrise, sunset and night-time passes are used to detect fire hotspots; information from afternoon passes are used for fire history and vegetation greenness images.

Image coverage

A satellite image is part of a continuous strip of the Earth seen by each satellite as it orbits; however, the areas covered by a single NOAA image and a single Landsat image are very different (Figure 7.2). A single image of the AVHRR sensor on a NOAA satellite is 2500 km wide and 4000 km long whereas a Landsat image is only 185 km x 185 km. Thus a single NOAA image can cover nearly two-thirds of Australia whereas it needs 400 Landsat scenes to cover all Australia.

Although each satellite progresses a similar distance between orbits to maintain its position relative to the sun, the size of the image influences the amount of overlap and the number of days between repetitive images. The NOAA satellites have a nine-day orbit cycle—a satellite passes directly overhead on every ninth day.

However, because the image is so wide, there is an overlap between images on successive days. This means that a specific location can usually be seen on five days of the nine-day orbit cycle although its position within the image, and the pixel size change as it drifts from one edge through the scene centre to the other edge.

In contrast, there is very little overlap between adjacent Landsat images; Landsat has a 16-day orbit cycle, and so any particular location can be observed only once every 16 days.

Image resolution

Pixel size is a function of the sensor's scan angle and its field of view (Figure 7.3). The ETM sensor on Landsat 7 has a very narrow scan angle and a narrow field of view, so all pixels are the same size across an image. The pixel size for a Landsat multispectral image (used to make colour pictures) is 30 m x 30 m, the panchromatic band (similar to black-and-white film) has a resolution of 15 m x 15 m and the thermal band (which is sensitive to heat) has a resolution of 60 m x 60 m.

The AVHRR sensor on NOAA satellites has a wide scan angle and a larger field of view. This combination, together with the curvature of the Earth, causes the pixel size to increase away from the centre of the image. For AVHRR images, the minimum pixel size directly underneath the satellite is 1.1 km x 1.1 km, but it stretches to 2.45 km x 6.7 km at the edge of the image.

Reflected sunlight, emitted energy and the detection of fires

Several channels of both the AVHRR and ETM sensors record the intensity of sunlight reflected from the Earth's surface (or from the clouds above it). The data are collected only during the day.

The data from the red and near-infrared regions of the spectrum are important for assessing vegetation because green leaves absorb red light but reflect near-infrared light. The greener the leaves the greater the contrast between the absorption and the reflection. As the leaves dry out during the dry season, the contrast between the red and near-infrared light decreases.

Figure 7.2 Comparison of area covered by NOAA and Landsat images

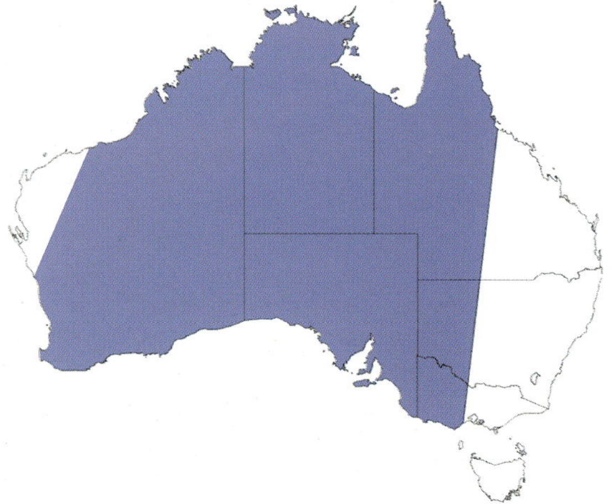

Single NOAA image covers two-thirds of continent.

400 Landsat scenes to cover Australia

After the vegetation has been burnt, black ash and char on the bare ground absorb most of the sunlight. These characteristics are used to provide the information for both the fire history maps and the vegetation greenness images.

The sensors also record emitted energy, which is a function of heat and temperature, and can do this during both day and night. This provides information for the detection of active fires or hotspots—that is, pixels which have a high temperature in contrast to surrounding pixels.

Hotspots are only determined from evening and night-time overpasses in the tropics. This restriction is because channel 3 of the AVHRR sensor (the most sensitive thermal channel) saturates at just 51°C—at soil temperatures typical in the tropics over much of the dry season. In turn, this presents a significant 'sampling' problem because many smaller fires go out at the end of the day and are undetected.

In addition, the minimum size or intensity of a grassfire that can be detected by hotspots is still unknown. However, hotspot detection can detect gas flares from oil wells in the Timor Sea and from the Mereenie Basin west of Alice Springs. The flares are hot enough to raise the average temperature of an entire 1 sq. km pixel.

The thermal channels from daytime NOAA and Landsat images are used in combination with the reflected channels to map the extent of fires. The black ash on the ground surface absorbs heat and raises the temperature of the burnt area above that of the surrounding unburnt landscape. This helps to improve the discrimination of fire extent.

The orbit cycle of each satellite determines the frequency of fire information updates. Fire history updates are restricted to the cycle of daytime overpasses. Therefore NOAA-derived fire maps can be updated on a nine-day cycle and Landsat-derived fire maps have a 16-day cycle.

The detection of hotspots from the NOAA satellites has fewer restrictions and updates are available daily, usually with two or three updates each day using different satellites. All the fire information updates require cloud-free conditions as no channels of either the AVHRR or ETM sensor can see through clouds. Only satellites with active sensors such as radar can do this, but these have yet to be tested for mapping burnt areas or detecting active fires.

Use of satellite data in fire monitoring

Two kinds of this satellite-based fire monitoring information are now available for land managers.

Active fire locations or 'hotspots'

The daily (or at least nightly) detection of fire hotspots provides remote land managers with a powerful source of information concerning the location of fires in any one region, or even in a large paddock. 'Near real time' information derived from hotspot images provides land managers with the opportunity to monitor the progress of approaching fires so that ongoing decisions regarding their management can be made.

Fire extent or fire history maps

These can provide land managers with a continuous record of the distribution and extent of fires throughout any one year on individual properties, or at regional and even continental scales.

Figure 7.3 Resolution of Landsat and NOAA images

Landsat image (left) provides a much clearer view of the fire extents than the AVHRR image (right)—there are nearly 1500 Landsat pixels within the smallest NOAA-AVHRR pixel.

For example, bushfire management agencies across northern Australia use these maps as monitoring and planning aids throughout the fire season. They help identify unburnt gaps in strategic fire control lines and can be used by land managers to develop a burning history for individual paddocks, or a fire history of fire-sensitive and surrounding vegetation types on a conservation reserve.

In combination with other data (e.g. known rates of fuel accumulation for different fuel types; climatic information), fire history maps can be used to assess fire risk.

NOAA-AVHRR and Landsat ETM satellite images are collected and processed by three agencies in three locations to provide Australia-wide coverage. The Department of Land Administration (DOLA) in Perth collects NOAA-AVHRR images for the western portion of Australia. The Queensland Department of Natural Resources and Mines in Brisbane (QDNRM) collects NOAA-AVHRR images for the eastern portion of Australia. Both agencies have websites for the distribution of active fire and fire extent information.

The Australian Centre for Remote Sensing (ACRES) is the agency responsible for Landsat images. It has receiving stations in Alice Springs and Hobart with processing based in Canberra. In addition, the Bureau of Meteorology offices in Melbourne and Darwin have satellite receiving stations for NOAA images and contribute to the Australia-wide coverage. The details of the websites for these agencies are provided at the end of this chapter.

Active fires or hotspots

What hotspot information is available?

Hotspot information is available as maps, lists of geographic coordinates and as digital files for use in a GIS.

The maps are picture files showing the location of active fires on a generalised land tenure map with some road and river information (Figure 7.4). The hotspots are colour-coded: red crosses have the highest probability of being active fires and meet all the criteria of their rating system; green crosses have a lower probability of being active fires and are labelled as possible fires.

The lists of geographic coordinates and the GIS digital files, as ArcView shapefiles (from QDNRM only), provide more accurate information on the location of hotspots. Users must be aware that DOLA provide fire locations in degrees and minutes of latitude and longitude whereas QDNRM use decimal degrees, so care has to be taken when comparing information from these two sources.

Figure 7.4 Hotspot map, Western Australia

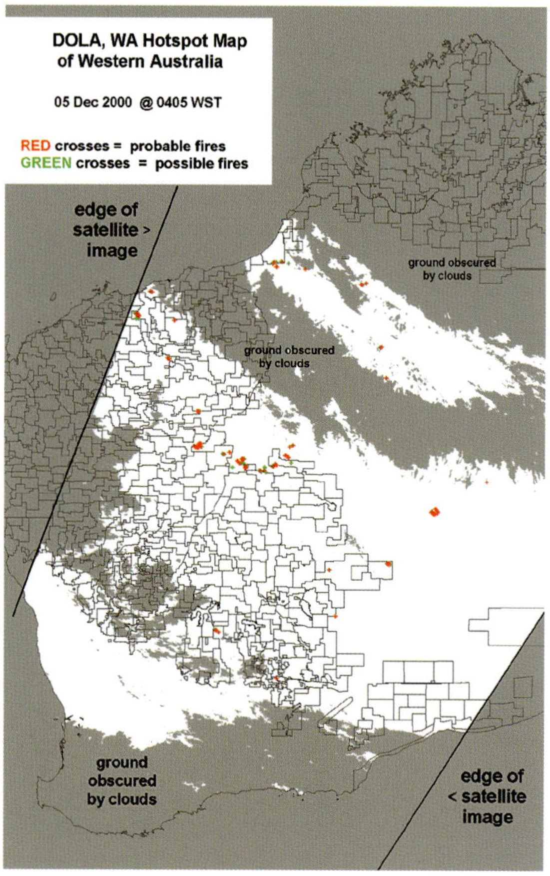

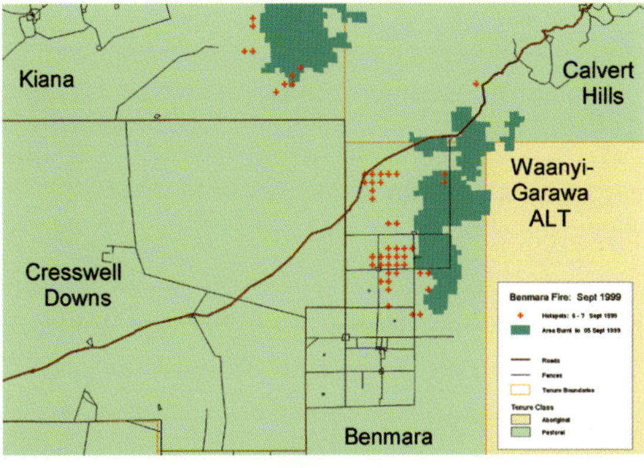

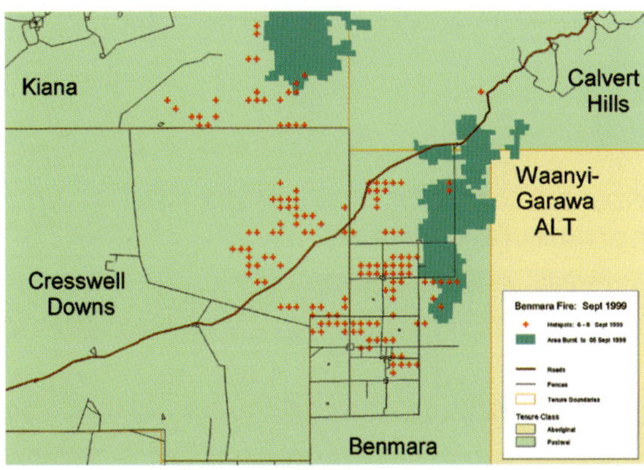

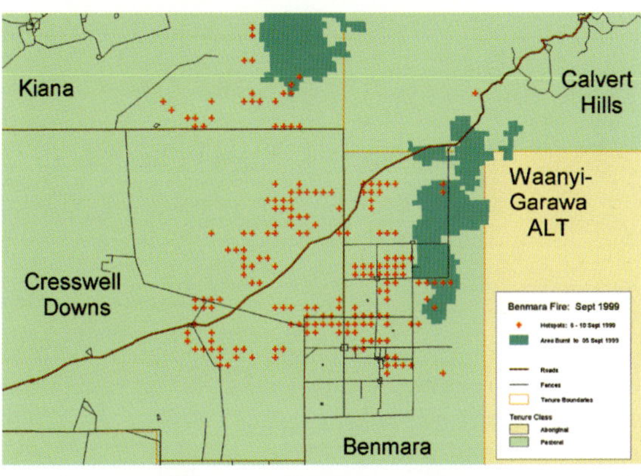

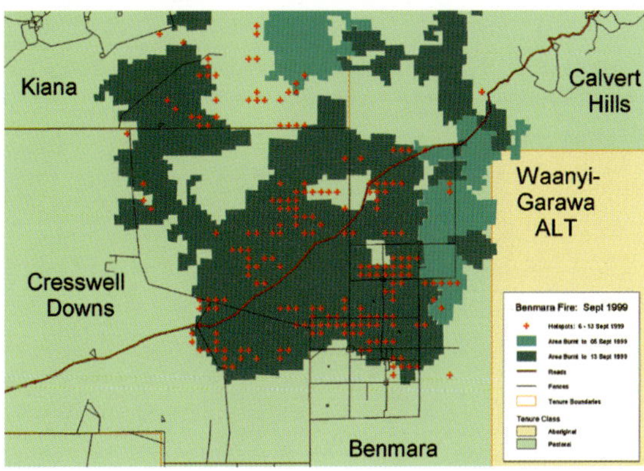

*The progression of hotspots (red dots) during the
interval between AVHRR fire history map updates*

The lists of hotspot coordinates can either be manually drawn on topographic maps or imported digitally into a GIS to display the hotspots overlaid on topographic or pastoral maps (or satellite images). The progress of a fire can be tracked using the list of geographic coordinates or the ArcView shapefiles to display a time sequence of hotspots.

Figure 7.5 shows a progression of hotspots. The first map shows the extent of the area burnt until 5 September 1999, as mapped from AVHRR images, plus hotspots for 6 and 7 September.

The second map shows the dramatic advance of the fire on 8 September with strong north-east winds. The third map shows the continued advance of the fire in a south-westerly direction during 9 and 10 September. Suppression activities by the pastoralist and Bushfire Council's officers restricted the continued spread of the fire and the fourth map shows the extent of the area burnt to 13 September.

This approach to disseminate hotspot information to station managers has been adopted as a standard within the NT. It relies on downloading the hotspot locations into a GIS. Within the GIS, the daily hotspot updates are plotted onto station or topographic maps with roads, rivers, fences and the extent of previously burnt areas. The daily maps are distributed to land managers by fax or email.

Limitations of NOAA hotspot data

The hotspot data have several limitations.

Delay

There is a several hour delay between the time that a satellite collects the data and when it becomes available on the website. This delay must be considered by the user in the context of weather conditions, fire behaviour and rate of fire spread.

Accuracy

The location accuracy of the hotspot data is, at best, ±1–2 km. This is associated with both the size of the image pixel and the quality of the image rectification, which is the process of transforming the diagonal line of the satellite orbit to the square pattern of maps in geographic coordinates. Pixel size varies with its position along each row of the image and is affected by the nine-day orbit cycle of the satellite.

Accuracy is best directly under the track of the satellite but decreases away from the centre of the image. The bigger the pixel, the less likely a fire will significantly increase the average temperature of that pixel to be identified as a hotspot. The best time for collecting data is when the satellite is looking straight down and the worst time is on days four and five of the cycle when the look angle is greatest. Thus the quality of the data are significantly lower on two out of every nine days.

The bad days are predicable but it depends on 'where you live'. Fortunately with four satellites available, there are few days when none of the satellites has a good view of any particular area.

Accuracy of the image rectification can be affected by the amount of cloud cover as clouds obscure ground features used to relate the image to a map.

Fire intensity

Typical daily fluctuations of fire intensity and the timing of the satellite data collection confound hotspot detection. Fires are usually less intense at night but increase again in the morning as the temperature and wind speeds increase. Many fires which die down and burn out in the evening (especially during the early dry season) will be missed.

Where can I get information?

Two websites—DOLA and QDNRM—provide daily information on the location of fire hotspots in northern Australia (see the list at the end of the chapter). QDNRM can provide automatic notification of hotspots by email if a pastoral land manager can provide four corner points for a rectangle of land surrounding the property. If any hotspots occur within the defined rectangle, an email is sent automatically listing the geographical coordinates of the hotspots, plus ArcView shapefiles.

Fire history

Fire extent (or fire history or fire scar) maps can be produced from either Landsat ETM or NOAA–AVHRR images.

NOAA–AVHRR

Fire history maps produced from NOAA–AVHRR imagery (Figure 7.6) have been available for the northern savanna regions of the NT and WA since 1993, and for all of northern Australia since 1997. The low cost, the large coverage of individual images and the opportunities for frequent updates combine to provide a valuable resource. The maps show the

Figure 7.6 AVHRR image of the north-west portion of the Top End

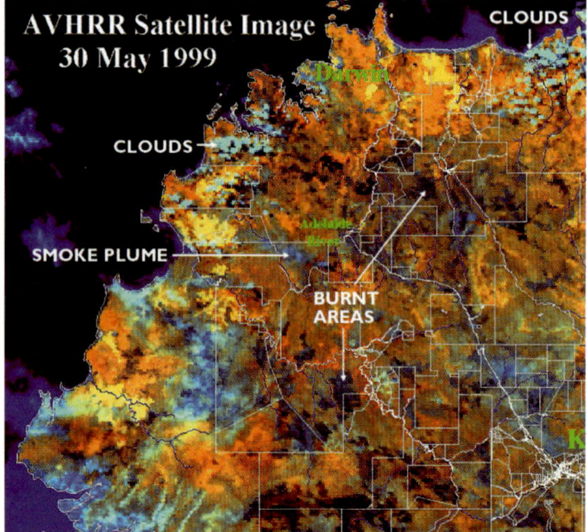

Recently burnt areas are black; clouds and smoke plumes may obscure fires.

Figure 7.7 Fire history map corresponding to the AVHRR image

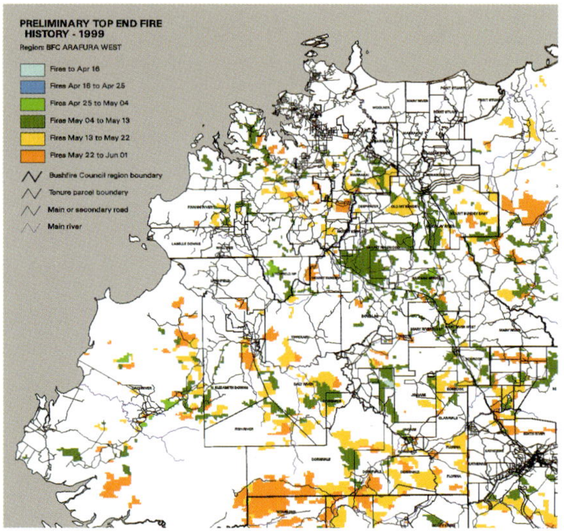

area burnt by fires colour-coded between different calendar dates (usually nine-day periods). Maps include an overlay of reference information, including the coastline, rivers, roads and land tenure (Figure 7.7).

Landsat

Due to higher imagery costs and small image areas, Landsat-based fire history maps are available only for small regional areas of northern Australia, typically for research purposes and management of national parks (Figure 4.3 p. 47). These maps have a high resolution (30 m x 30 m minimum mapping area) but updates are prepared from only three images per year representing early, middle and late dry season fires.

Fire history data for the tropical savannas are accumulated throughout the dry season (April to November–December). Plant growth during the wet resets the vegetation patterns and masks the previous fire patterns; fire history maps begin with a clean sheet at the start of the dry season.

Where can I get fire history maps?

Fire history information is available as either maps or GIS data files from the DOLA website. The maps are available for all of WA and the NT, or for regional areas (Figures 7.8 and 7.9).

Are printed fire history maps available?

Several standard fire history maps are produced and distributed—mainly to regional fire control officers and to managers of national parks. Maps can also be produced for pastoral properties or other land tenure units by request to either DOLA or Bushfires Council of the NT (Figure 7.10). These plots include the summary statistics of the area burnt by each image date calculated as both the area extent in sq. km and as a proportion of the total area of the tenure unit.

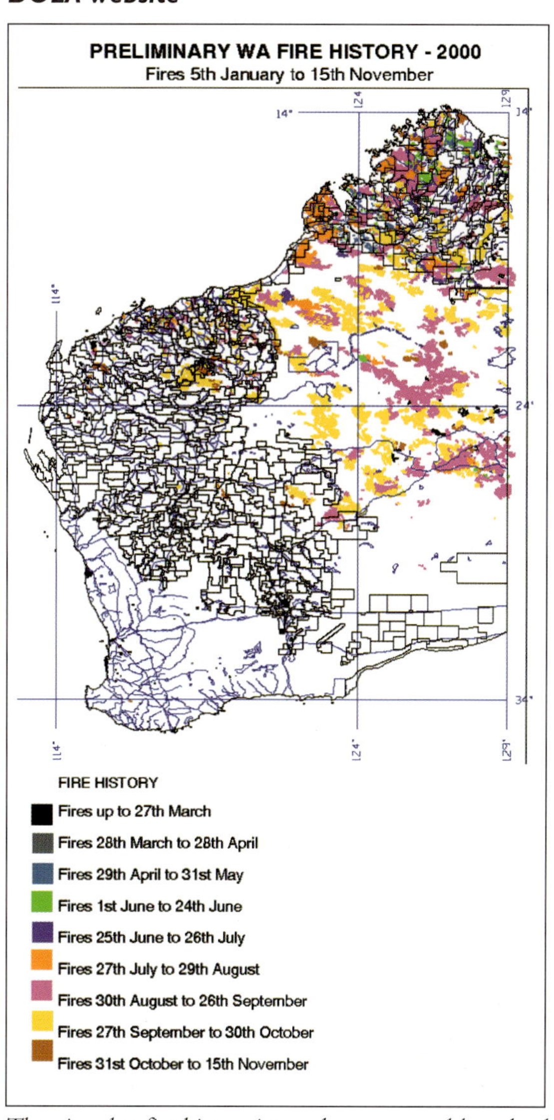

Figure 7.8 Fire history map, Western Australia, DOLA website

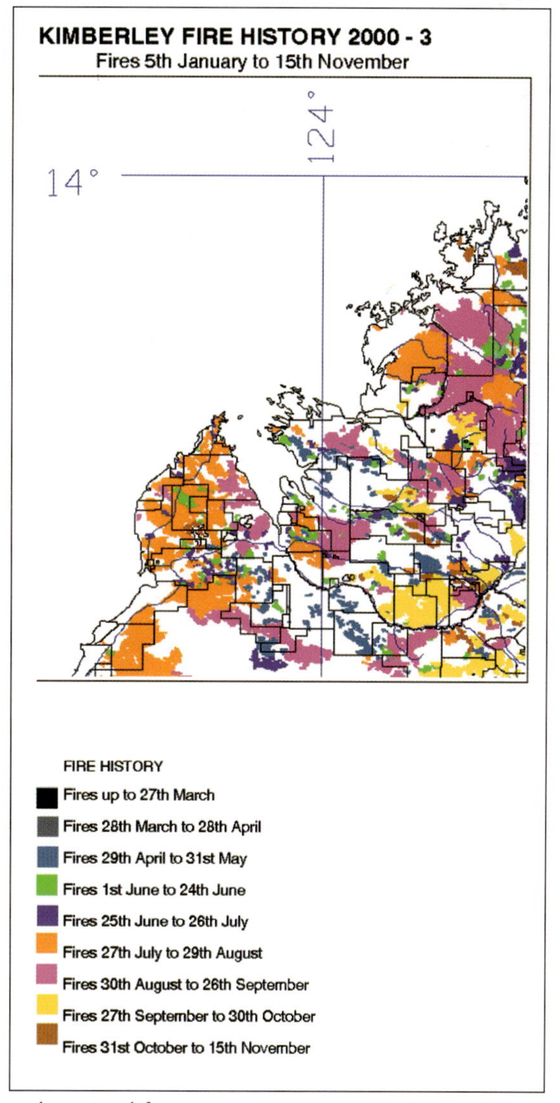

Figure 7.9 Enlargement, Derby–Broome region of WA

The nine-day fire history intervals are grouped by calendar month to simplify interpretation

The historical sequence of fire history maps provides an insight into the different fire regimes of northern Australia. The information provides a visual comparison between years, or across regions and land tenures as graphs (Figures 7.11 and 7.12) or maps. Combining all the fire history maps from 1993 to 2000 creates maps of fire frequency and fire interval. Fire frequency maps can indicate the number of times each location has been burnt during the past eight years (Figure 7.13) whereas fire interval maps show the time intervals between two successive fires.

Figure 7.10 A 1996 fire history map for a grazing property with 17 fire history intervals

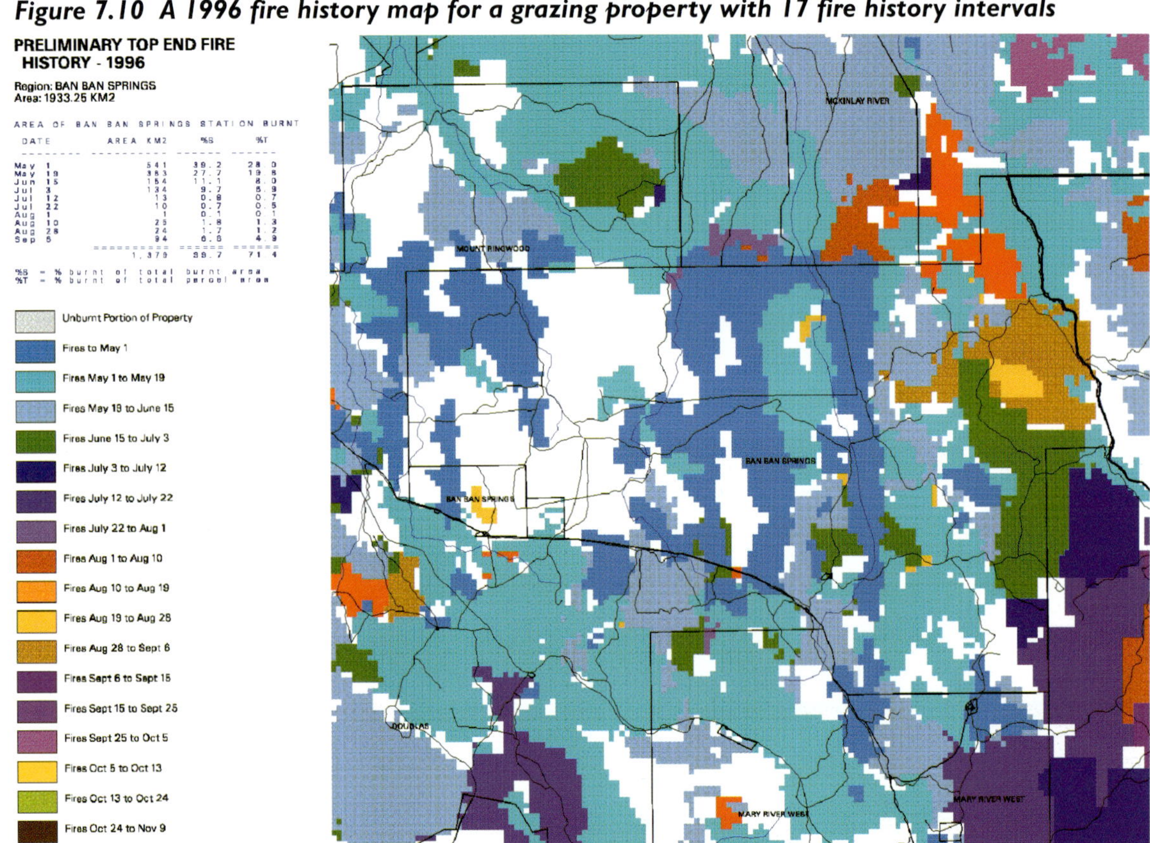

Figure 7.11 Fire history, bar graph

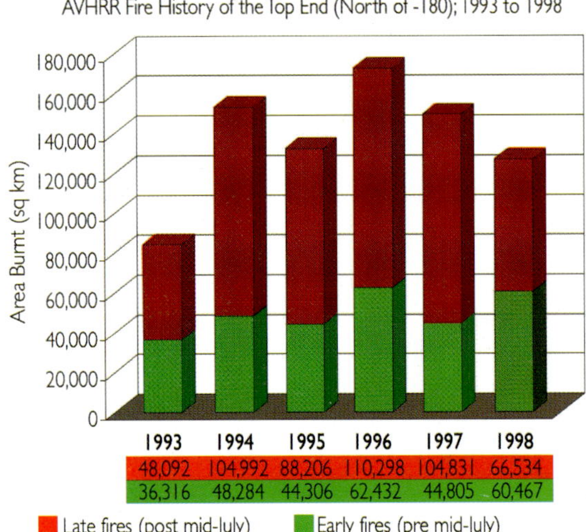

Fire history data describing area burnt across the whole Top End of the NT for 1993 to 1998. The columns are split for early fires (before mid-July) and late fires (after mid-July). Total area burnt was highest in 1996.

Figure 7.12 Fire history, line graph

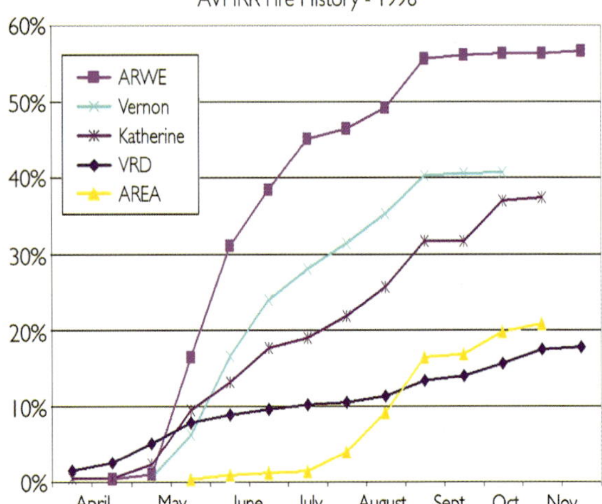

Fire history data showing timing of fires. Regions with more intensive land use are burnt early (Arafura Region West—ARWE); Arnhem Land (Arafura Region East—AREA) is burnt mainly by large wildfires late in the dry season.

Is other information available in the maps?

Additional information can be included with the maps on request. For example, Bushfires Council officers record the location of their ACB lines using a GPS on board the aircraft; having these lines on a map allows them to assess their effectiveness against the fires detected by the satellite (Figure 7.14). Other information options include infrastructure, such as fence lines and bores, or topographic maps as background images.

Figure 7.14 Fire history, Katherine

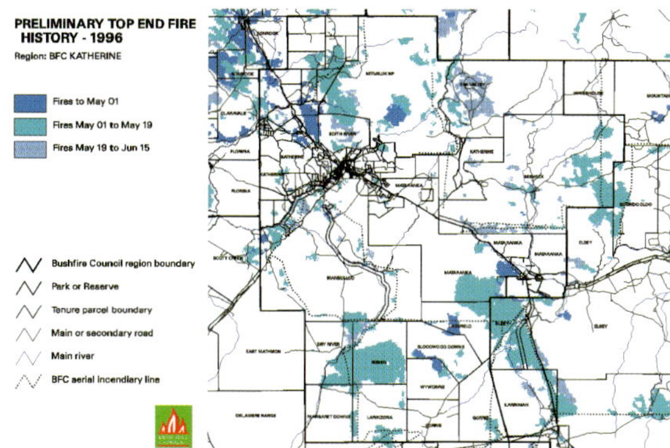

A fire history map of the Katherine BFC region showing ACB lines and early fires associated with a burning program

What are the limitations of the fire history maps?

The usefulness of fire history maps is limited by clouds, small fires and wet season burns.

Clouds

The sensors cannot see through clouds. Unfortunately clouds are frequent in the late dry season months of October and November when there are many large wildfires.

Small fires

The AVHRR sensor has a spatial resolution of 1 sq. km. The minimum size of a fire which can be mapped will depend on its intensity and shape, and on the type of vegetation. To be readily identified a fire must burn a strip or a block over a portion of an area of 3–5 sq. km; many small fires in the early dry season are not detected.

Wet season burns

Wet season burns are rarely mapped as part of the fire history mapping. The fires are generally small and patchy, at a time with few cloud-free days, and are masked by regrowth by the end of the wet season.

Figure 7.13 1993 to 2000 fire frequency map of the Top End, NT

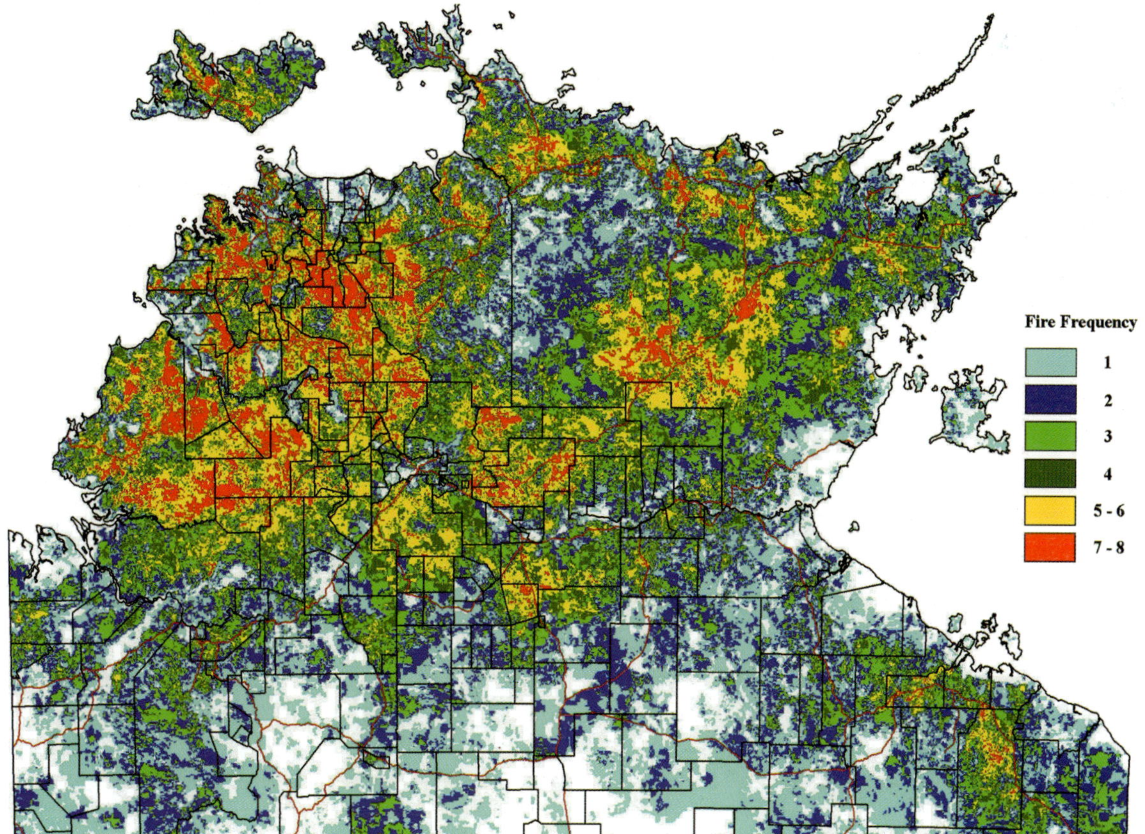

Monitoring vegetation greenness

Across northern Australia there is considerable annual variation in the rate of curing (or drying), curing state and fuel moisture content between major plant communities. There is also significant variation in fuel curing state within the same communities but in different localities throughout a region depending on the amount and distribution of rainfall during the preceding wet season. Having access to accurate and up-to-date curing information allows regional fire authorities and land managers to undertake more effective regional and property fire management. The daytime AVHRR satellite images can be used to determine vegetation 'greenness' or, conversely, the state of curing.

A knowledge of the curing rate is important to fire management agencies and land managers. Grass fuels grow and cure very rapidly. The window of opportunity for safe and effective hazard reduction burning over most savanna regions remains relatively small. It is also important to be able to accurately identify the period during the dry season when wildfire risk increases to critical levels. Furthermore, the adoption of prescribed burning practices implemented as part of normal grazing management relies on managers being able to utilise optimum fuel conditions suited to the objectives of fire management.

The large variation in fuel curing across the landscape throughout the year can make these tasks very difficult. Some pasture communities, such as annual sorghum on sandy soils, will cure rapidly and present a fire risk much earlier than others, such as perennial grasses on cracking clays. A poor or patchy wet season may result in fuel rapidly becoming fully cured in some areas while adjacent areas remain green.

This variation in fuel curing often hampers the effectiveness of strategic aerial control burning programs where, to minimise costs, large areas are burnt rapidly from fixed-wing aircraft.

How is vegetation greenness measured and calculated?

Vegetation greenness is based on the characteristics of sunlight reflected from plant leaves. This greenness value is also known as the Normalised Difference Vegetation Index (NDVI), which is calculated from AVHRR channels 1 and 2 on the NOAA-B satellite. Green leaves absorb red light recorded by AVHRR channel 1, but reflect near-infrared light recorded by AVHRR channel 2. The result is a high NDVI value. As leaves dry out, they absorb less red light and reflect less near-infrared light and the NDVI value decreases. NDVI values can be derived for the whole area of an AVHRR image.

The images portray the amount of green, actively growing vegetation within each pixel as a relative measure (Figure 7.15). As the dry season progresses, successive NDVI images respond to the progressive drying out of vegetation (less greenness). When the rains commence again, NDVI images can help identify those areas which are greening up again in response to localised rainstorms. The direct relationship between the vegetation greennness values and the amount of green vegetation on the ground is in the process of being verified, both in the Top End of the NT and the Kimberley.

Figure 7.15 Vegetation greenness

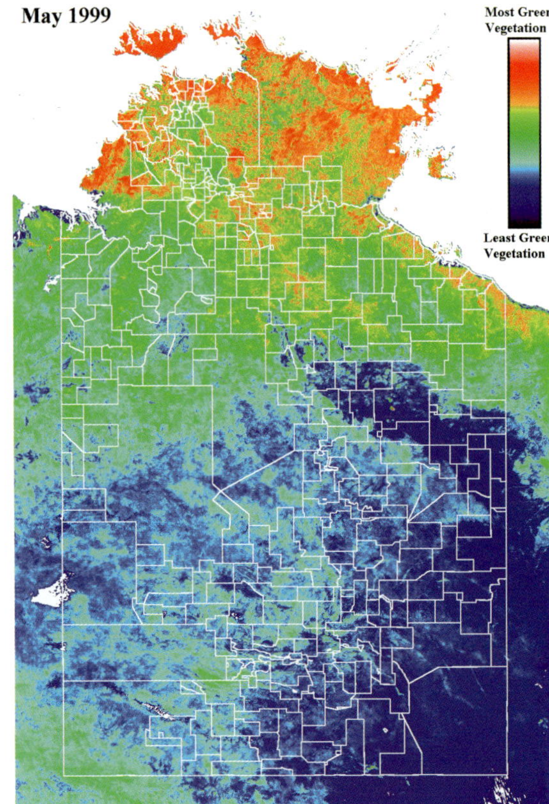

Vegetation greenness image of the NT. Blue indicates nil or dry vegetation, progressing through green and red to white.

Fuel loads, curing and moisture contents of a pature as the dry season progresses

February—2500 kg/ha fuel, 20% cured, 65% moisture content

March—4000 kg/ha fuel, 25% cured, 60% moisture content

May—3000 kg/ha fuel, 65% cured, 20% moisture content

July—000 kg/ha fuel, 80% cured, 20% moisture content

How is curing related to greenness?

The converse of vegetation greenness is vegetation curing. Curing is the loss of greenness and moisture in vegetation as it dries and dies off after the wet season growing period (Figure 7.16). It is expressed as the proportion of dead (brown) material to live material.

Fuel curing refers to the relative 'greenness' of the fuel; it affects fuel flammability and the potential rate of fire spread.

Where are greenness images available?

The vegetation greenness images are prepared bimonthly (24 times a year) by DOLA in Perth. Each image combines information from numerous satellite overpasses within each half-monthly period to eliminate cloud-affected areas and maximise the greenness value within each pixel. The final compilation is generally available within a week of the end of each collection period.

The most important aspect of the information is the frequency of its availability and the opportunity it provides to follow patterns through time and for making comparisons between seasons. The current archive of data for WA and NT begins in January 1993. An archive of monthly NDVI images of Australia are available through the Northern Territory Department of Primary Industry and Fisheries and the Bureau of Meteorology websites.

What else can vegetation greenness images tell us?

Besides showing the state of growth or curing of vegetation, greenness images can be interpreted to provide:

The rate of curing

The sequence of fortnightly images show the differential rates of change of curing within different vegetation and soil types (Figure 7.17).

The geographical extent of the wet season

The significance of variability between wet seasons becomes more important towards the arid interior. In the parts of northern Australia where rain gauges are widely separated, vegetation greenness images complement rainfall data across the whole landscape (Figure 7.18).

The start of the wet season

Vegetation responds as humidity increases before the rains begin. The vegetation greenness images are sensitive to the flush of new growth, primarily by the trees and shrubs of the open woodlands.

The length of the wet season

As the vegetation dries out after the end of the wet, the 'green bits' show where the wet season persisted. This can affect management programs such as aerial control burning.

What are the limitations of the vegetation greenness images?

As the vegetation greenness characteristic of each vegetation community on each soil type is slightly different, images are relative, rather than absolute. A vegetation greenness value of 0.4 in a woodland community on red soil does not equate to a value of 0.4 in open Mitchell grass plains on black soil. Nevertheless, land managers familiar with their country and its variability will find vegetation greenness images to be a useful management tool for interpreting the moisture status of different vegetation types and their associated soils.

Figure 7.17 Vegetation greenness, April 1995–March 1996

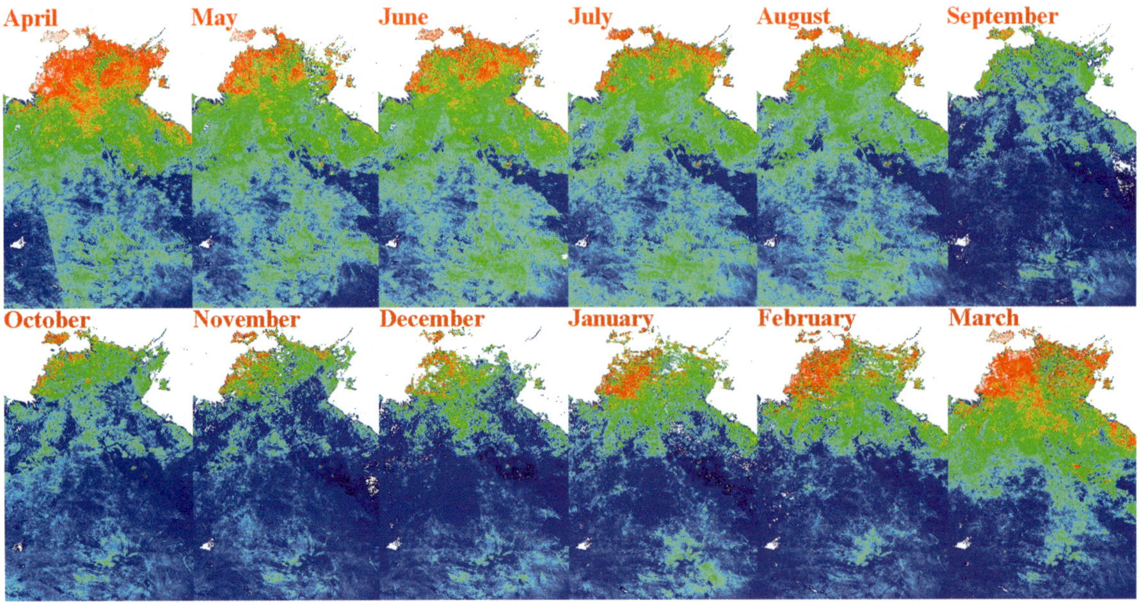

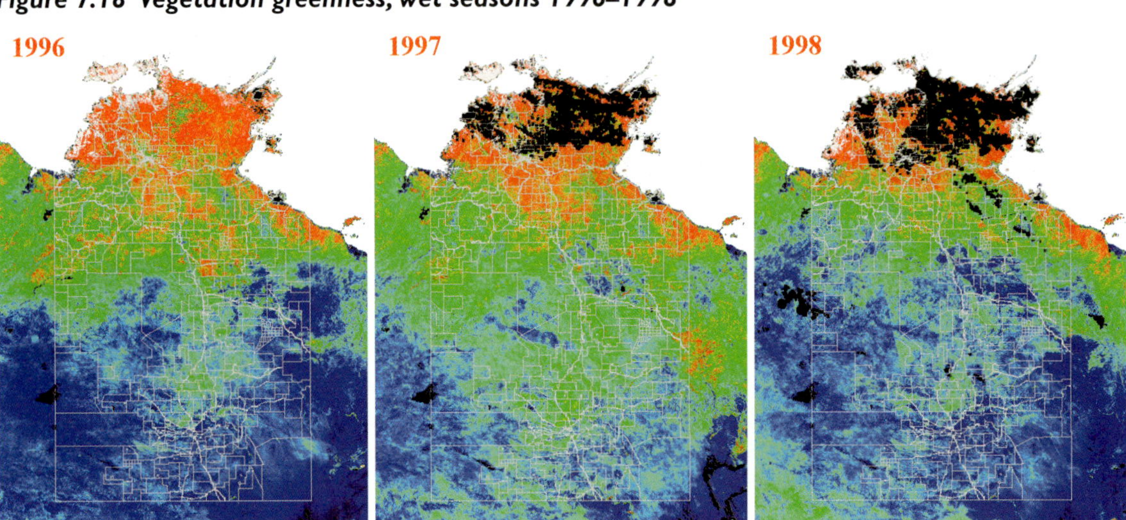

A sequence of vegetation greenness images for the Top End between April 1995 and March 1996 showing curing as the dry season progresses, then greening up over the next wet season

Figure 7.18 Vegetation greenness, wet seasons 1996–1998

Differences between the extents of the wet seasons from 1996 to 1998 as gauged from images of maximum vegetation greenness for each year

Modelling seasons and fuel conditions

Most rural fire agencies and land managers are aware that the effectiveness and safety of prescribed burns, and the risk of wildfire, are influenced by the amount and curing state of fuel. Large variation in fuel conditions can occur between seasons (temporal variation) and throughout the landscape (spatial variation).

Sequences of above-average seasonal rainfall and low grazing pressure result in the accumulation of fuel and the increased risk of wildfires. Alternatively, periods of drought or overgrazing reduce grassy fuels, often eliminating any possibility of fire. Similarly, rainfall distribution affects pasture greenness, the curing state of fuel and its susceptibility to burning throughout the year and across the landscape.

It is now possible to provide rural fire agencies and land managers with reliable estimates of seasonal rainfall, fuel curing, fuel loads and fire risk through computer simulation—or modelling (Figures 7.19–7.23). This is now being done by the Aussie GRASS Project (the Australian Grassland and Rangeland Assessment by Spatial Simulation Project).

Aussie GRASS is a national collaborative research project that aims to develop a spatial modelling framework which allows the condition of Australia's grazing lands to be assessed and monitored. Each month, the Aussie GRASS computer model simulates the pasture growth and fuel conditions across Australia by dividing it up into around 250,000 5-km grid cells. In each grid cell, the model integrates daily rainfall and climate data (temperature, radiation, vapour pressure and evaporation) received from the Bureau of Meteorology, along with soil type and pasture community parameters, tree basal area and livestock numbers.

Aussie GRASS products can assist with planning burning operations and can assess fire risk at scales from large properties to whole regions. Estimates of Total Standing Dry Matter (TSDM), an indicator of fuel load, use computer calculations of pasture carry-over, pasture growth and livestock grazing. The curing index is based on the proportion of green and dead grass output from the model, while the grassfire risk is determined from both fuel load (TSDM) and curing index.

Figure 7.19 Rainfall relative to historical records, November 2000–April 2001

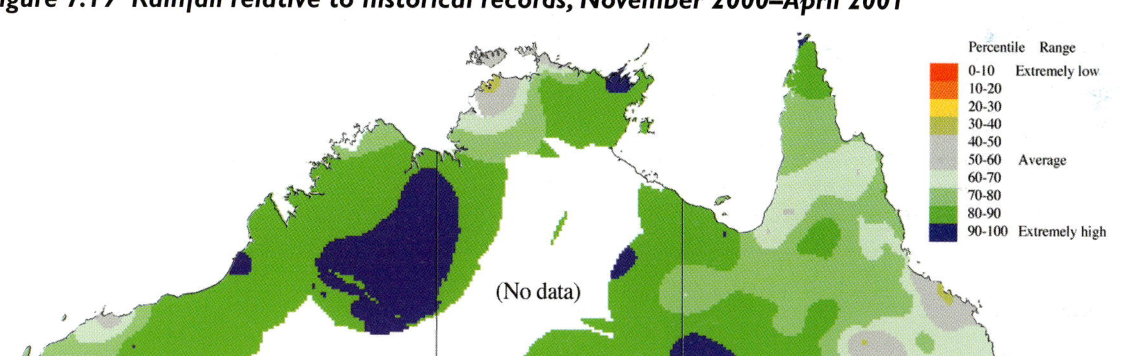

Figure 7.20 Total standing dry matter (kg DM/ha), April 2001 (Experimental prototype)

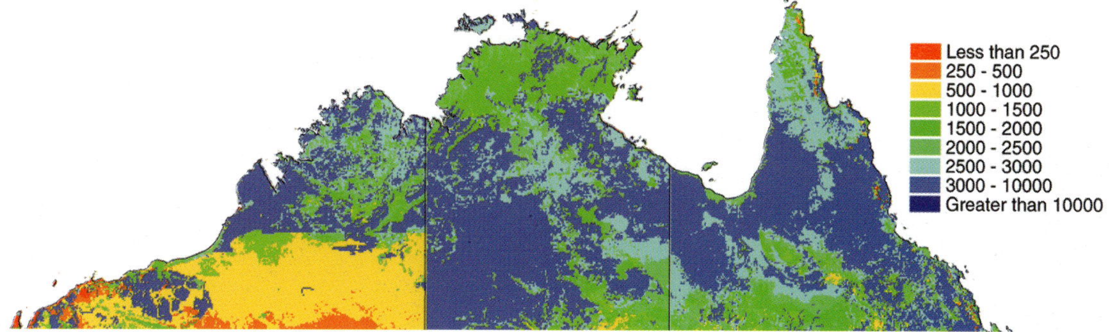

A range of Aussie GRASS products including relative seasonal rainfall, actual total TSDM, relative TSDM, curing index, and grassland fire risk can be downloaded from the internet

Aussie GRASS model outputs are often presented in a ranked or percentile format making it easy to compare fire risk between seasons. Because the Aussie GRASS model outputs rely on large data sets with varying levels of resolution while livestock numbers are available only at regional scales, some products, such as TSDM, should be regarded as estimates only.

Experimental Aussie GRASS map products are updated monthly and can be viewed and downloaded directly from the internet. For help on how to access this information, contact the Aussie GRASS coordinator or Queensland Department of Natural Resources and Mines on (07) 3896 9502.

Further reading

Carter, J. O., Hall, W. B., Brook, K. D., Day, K. A. and Paull, C. J. (1998). Aussie GRASS: Australian simulation and rangeland assessment by spatial simulation. *Applications of Seasonal Climate Forecasting in Agricultural and Natural Ecosystems—The Australian Experience.* (Eds G. Hammer, N. Nicholls and C. Mitchell.) Kluwer Academic Press, Netherlands. pp. 329–349 .

Figure 7.21 Total standing dry matter relative to historical records, April 2001 (Experimental prototype)

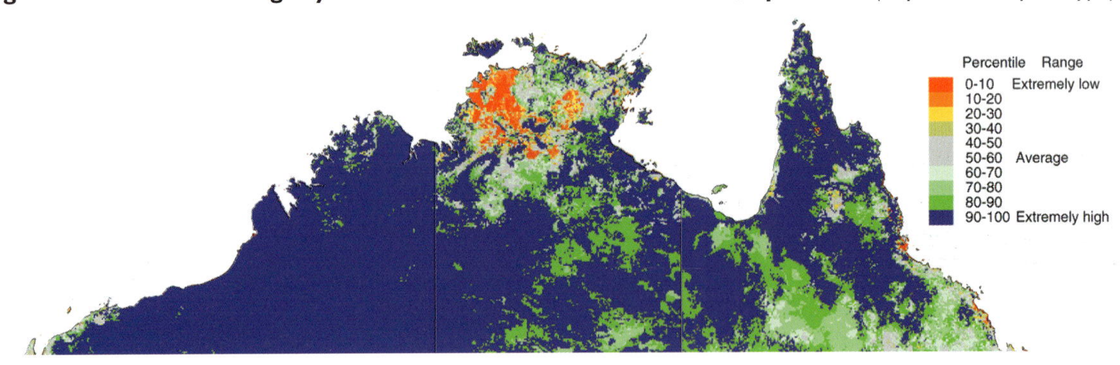

Figure 7.22 Potential grassfire risk, April 2001 (Experimental prototype)

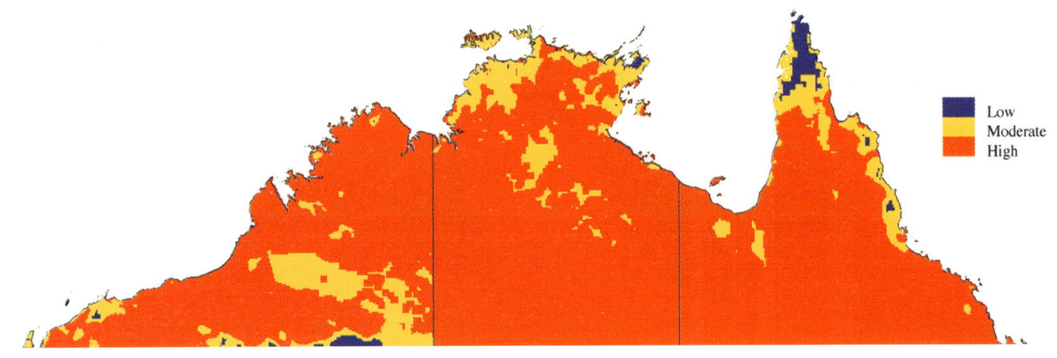

Figure 7.23 Curing index, April 2001 (Experimental prototype)

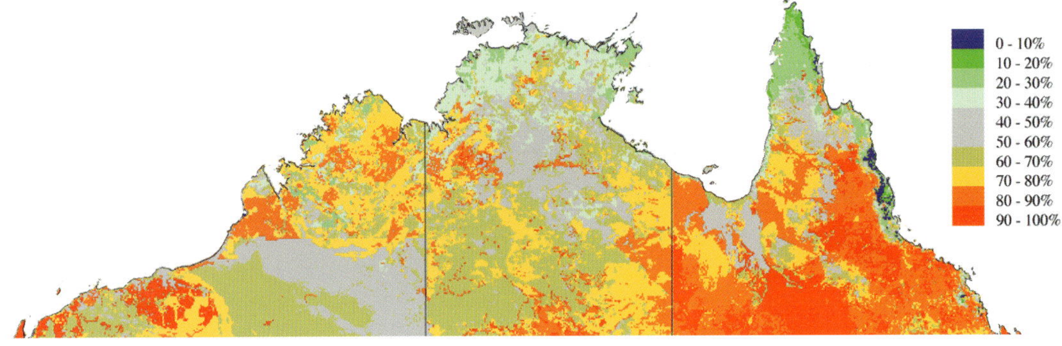

Satellite monitoring websites

Hotspot Information
Two websites provide daily information on the location of fire hotspots in northern Australia.

DOLA: http://www.rss.dola.wa.gov.au

QDNRM: http://www.dnr.qld.gov.au/longpdk/ SatelliteFireMonitor

Fire history and fire information websites
Bushfires Council of the NT:
http://www.nt.gov.au/bfc

- information on the Bushfires Council, including links to other websites.

DOLA (Department of Land Administration, Perth WA): http://www.rss.dola.gov.au

- provides fire-related satellite information for WA and the NT.

Vegetation greenness websites
DOLA: http://www.rss.dola.wa.gov.au

NTDPIF: http://www.nt.gov.au/dpif/rangelands

- information related to the NT pastoral industry.

Bureau of Meteorology: http://www.bom.gov.au

- greenness images of Australia from January 1997.

Fire ecology websites
Tropical Savannas Cooperative Research Centre, Darwin: http://savanna.ntu.edu.au

- overview of research programs in northern Australia including a demonstration of a GIS system for the Victoria River Downs district of the NT.

CSIRO Tropical Ecology Research Centre, Darwin: http://www.dar.dwe.csiro.au

- overview of their research programs including a series of fire-related information sheets.

Australian National University, Canberra: http://online.anu.edu.au/Forestry/fire/FNET/ introtofirenet.html

- Fire-Net 'the on-line information service for everyone interested in rural and landscape fire'.

Remote sensing information
Further information on the technical characteristics of these and other satellites is available from:

ACRES: http://www.auslig.gov.au

NASA: http://www.nasa.com

DOLA: http://www.rss.dola.wa.gov.au

SPOT: http://www.spotimage.com

Landsat: http://landsat7.usgs.gov

http://landsat.gsfc.nasa.gov/main/documentation.html

http://ltpwww.gsfc.nasa.gov/IAS/handbook/ handbook_toc.html

NOAA-AVHRR: http://poes2.gsfc.nasa.gov/ history/history_home.htm

http://www.saa.noaa.gov/

Tutorials
http://rst.gsfc.nasa.gov

http://www.ccrs.nrcan.gc.ca/ccrs/eduref/tutorial/ tutore.html

http://www.ccrs.nrcan.gc.ca/ccrs/eduref/youthkit/ edukite.html

The future

Over coming decades, north Australian savannas will be affected by major global trends including advancing technology, increasing populations, globalisation of markets, increasing carbon dioxide concentrations and changing climates. Locally, increasing pressure is being applied to land managers to demonstrate ecological sustainability. At the same time, multiple land uses are emerging in this region where pastoralism has dominated in many areas for the past century. Land tenure issues to do with native title rights are still being resolved, and on many properties land management goals are shifting to incorporate aspects of commercial pastoralism, tourism and traditional Aboriginal land use. In this chapter we will look at some of these trends in more detail to see how they will affect fire management in Australia's tropical savannas.

Greenhouse issues and carbon trading

By the end of the 21st century the atmospheric concentrations of carbon dioxide and other greenhouse gases will most likely be double their pre-industrial levels. These changes are likely to affect climates and may increase sea-levels around the world. In north Australia there has been a general increase in heavy rainfall over the past century. This trend is consistent with the predictions from climate models under enhanced greenhouse conditions. The frequency of extreme events such as floods and droughts will probably increase. The best predictions for precipitation are that wet season rainfall in north Australia could decline by up to 8%. By the year 2030, temperatures are likely to increase by 0.4–1.4°C inland and by 0.3–1.0°C in coastal regions of north Australia. (See Australia's Second National Report under the United Nations Framework Convention on Climate Change: http://www.greenhouse.gov.au/)

As well as affecting climate the increasing carbon dioxide concentrations are likely to lead to increased plant growth. Trees and shrubs will probably be favoured more than tropical grasses. However, increasing temperatures may offset some of these gains. For example, the dominance of Mitchell grass could be put at risk in many areas by even a 1°C rise. The increased demands for water by plants, coupled with decreased rainfall and possibly fewer rain days, could increase water stress on plants leading to north Australian ecosystems becoming more arid.

Higher temperatures may also cause fuels to become drier thus the rates of fire spread may increase with a resultant rise in fire intensities. This increase in intensities would be about 2% for each 1°C rise in temperature. However, possible changes in wind speeds, for which we do not have good predictions, would have a much greater affect. An increase in average wind speeds of 1 km/h could increase fire intensities by up to 20%.

Around 30% of Australia's human-induced greenhouse gas emissions are from land-management based activities including cropping, grazing, land clearing and forestry. Worldwide, fires lit by people in savannas are also important sources of greenhouse gases that affect the atmosphere and climate. In Australia, greenhouse gases from burning of savannas and grasslands comprise about 2% of the total human-sourced emissions of these gases (Table 8.1).

Most of the savanna burning comes from north Australia with Queensland, Northern Territory and Western Australia accounting for more than 98% (Figure 8.1). The figures for Western Australia include substantial burning in the Pilbara and Hamersley Ranges outside of the tropical savannas.

Table 8.1 The relative contribution of savanna burning to Australian emissions of greenhouse gases in 1996. CO$_2$-equivalent emissions (Gg). *Under the current greenhouse protocol CO$_2$ is not seen as a significant net emission from savanna fires as virtually all CO$_2$ emitted by fires is considered to be reabsorbed by vegetation regrowth.*

Source	CH$_4$	N$_2$O	Perfluorocarbons	Total	% total
Savanna burning	5707	4319	0	10,026	2.4
Australian total	109,215	24,192	1484	419,217	100

Source: National Greenhouse Gas Inventory Committee (1998), National greenhouse gas inventory 1996, Australian Greenhouse Office, Canberra.

In order to reduce the impacts of the enhanced greenhouse effect, a range of international initiatives are designed to reduce greenhouse gas emissions and increase carbon storage (see http://www. greenhouse.gov.au). The impacts of these initiatives on savanna burning are yet to be worked out. Frequent fires keep the density of trees and consequently the amount of carbon stored in savannas below what is possible for particular climates and soil types. In many pastoral regions the exclusion of fires has lead to increased thickening of woody vegetation, while in other areas very frequent fires are suppressing tree growth. In the future, carbon trading could allow the carbon storage in trees to have a commercial value that might offset reductions in pastoral productivity resulting from tree thickening. However, there are still substantial scientific issues to be clarified and agreed on before carbon trading could be a realistic source of income for tropical savanna land. The challenge will be to develop burning strategies to optimise potential economic gains from increasing carbon storage as well as from sound pasture management. It will also be a challenge to manage for biodiversity with burning if carbon storage is in the mix of land management goals. For further information on climate change issues and the greenhouse effect see http://www.ipcc.ch/.

Figure 8.1 Greenhouse gas contributions

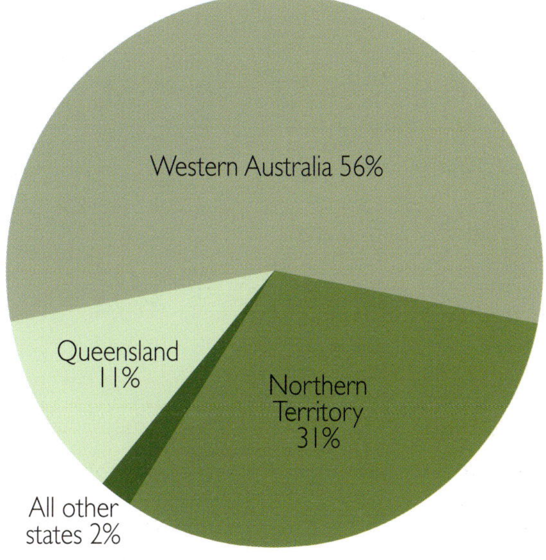

The relative contribution of each northern state to all of Australia's greenhouse gasses emitted by savanna and grassland burning. Source: National Greenhouse Gas Inventory Committee (1988), National greenhouse inventory 1996, Australian Greenhouse Office, Canberra

Globalisation of markets

In general, rangeland ecosystems are unable to provide high volumes of livestock and the supply of stock is highly dependant on climatic variability. Rangeland produce is usually strongly affected by global market forces but unable to influence them. Over the next decade, globalisation of economies is likely to lead to a continued downward pressure on prices of agricultural commodities. In northern Australia, live cattle exports are important regionally, but changes in beef production in Europe or the Americas could have major impacts on Australian production.

Developing niche markets is one possibility to ensure continued economic viability. Taking advantage of sustainable production systems and beef that has been produced without chemical additives could help ensure sustained market access and market advantage for beef from north Australian savannas.

Some system for demonstrating sustainability would be required to gain this market advantage. Adopting accredited environmental management systems such as ISO 14000 is one possibility. This international standard requires that the potential impacts of management activities are considered and steps taken to minimise those impacts. Protocols to monitor the impacts and continually improve management to reduce those impacts are also required. Fire, being one of the cheapest land management tools in north Australia, would need to be considered within any such framework. Clearly, the research into fire impacts and the monitoring systems already in place and being developed will be critical (see Chapters 4 and 7).

Live cattle export

Biodiversity

The *Commonwealth Environment Protection and Biodiversity Conservation Act 1999* aims to promote the conservation of biodiversity through cooperative approaches to the protection and management of the environment involving governments, the community, landholders and indigenous peoples. Land managers will need to deal with issues of biodiversity conservation, but there are many gaps in knowledge about how pastoralism and fire interact in their effects on biodiversity of Australian savannas. Further research will be needed to provide information to ensure that land management practices allow natural ecosystems to be sustained. However, information alone will not ensure these ecosystems are sustained—there may need to be policy, economic and cultural changes as well. Monitoring will play a key role here and, in the future, land managers may need to play a larger role in monitoring biodiversity as part of an adaptive management framework.

Air quality

In 1998, all Australian states and territories agreed on a set of national air quality standards. These assist in ensuring that all Australians enjoy the benefit of equivalent protection from air pollution. Because smoke from fuel reduction burning and bushfires can have wide regional impacts and affect communities far removed from the burn site, these standards may impact on burning practices.

Bushfire smoke affects human health mainly because of inhalation of tiny particles that increase the prevalence of a range of respiratory and cardio-vascular conditions in people exposed to them. The current air quality standard allows a limit of five days each year when more than 50 g of such particles can occur in one cubic metre of air.

Experience in Sydney suggests that these standards should not unduly affect fire management. There, in 1997, fine particle concentrations were measured when smoke from prescribed burning covered the city for three days. Despite the persistence of this highly visible blanket of smoke, the ambient air across Sydney did not exceed the new Australian standard. Although this is only one example, it shows that hazard reduction burning can still be carried out without breaching standards of air quality.

Biodiversity is influenced by savanna fire regimes.

Fire approaches suburbs in Palmerston near Darwin. Bushfire smoke affects human health.

The damaging effects of savanna wildfires often impacts urban areas, Northern Territory News.

There are however, no known threshold levels of particles below which impacts on human health cannot be detected—even very low concentrations may cause some problems. Because of this, a common philosophy taken by environmental authorities is that emissions of particles should be reduced to the lowest extent feasible. Land managers proposing to use fire as a tool to protect life and property and manage the environment should therefore make every effort to consider the potential health impacts of their activities. In particular, the redistribution of smoke due to atmospheric circulation patterns needs to be taken into account in order to limit the potential impact of smoke on large population centres.

For further information, see the National Environment Protection Measure (NEPM) for ambient air quality http://www.nepc.gov.au/air/air_nepm.6.98.html.

An appropriate management system

Our ultimate goal must be to develop fire regimes that give the best possible outcomes for society's diverse aims in land management. Appropriate fire regimes will differ across environments and land uses of northern Australia, but, to varying degrees, all the issues described and many others will need to be taken into account.

Clearly, this ambitious goal represents a major challenge to land managers, researchers and legislators of north Australia. Adaptive management principles will be crucial (Figure 8.2). The primary questions are 'What do we want from our environment?' What do we want for tree density, pasture quality, conservation status, the air about us?

These questions need to be asked and answered at various scales ranging from small patches of vegetation, to catchments, regions and the whole of north Australia. No simple answer will address all these concerns across all these scales, but rather a process is needed that will continually refine management and ensures that it meets the diverse and gradually changing goals of the broader society. Clearly, good communication across a social network will be key to this process.

The next step is determining how fire can help to achieve those land management goals. What regime of fire frequency, timing, intensity and patchiness should we aim at? How will these vary from patch to patch, property to property, region to region, industry to industry? Much is already known about short- and long-term effects of fire but there are still many gaps in the knowledge. Once a suitable fire regime has been decided upon, appropriate on-ground fire management techniques need to be applied. These must take into account the economics of fire management. We then need to monitor both the fires themselves and their impacts and provide feedback for better fire management.

Are we achieving the fire regimes we want? Are they having the environmental effects we predicted? What do we need to change? Again, good communication will be essential to ensuring that not only is up-to-date information available to help this process, but everyone with a legitimate interest in the outcomes can participate in the decisions.

Figure 8.2 Fire strategies in the context of land management

Native title and fire management

In 1992, in what has become known as the Mabo decision, the High Court recognised native title rights of indigenous people. Those rights derived from a connection to land based on traditional laws and customs, including the use of resources or management of land in accordance with tradition. The *Native Title Act*, proclaimed in 1993, gave statutory effect to Mabo, but provided for alienation of native title where rights had previously been granted to others. However, in the subsequent Wik decision (1996), the High Court ruled that the grant of pastoral leases did not necessarily extinguish native title. The Court determined that some rights could coexist with a pastoral lease issued under statute (as distinct from under the superseded Royal prerogative). Where the statute or instrument provided a specific right to the holder of a title, this right prevailed over any conflicting native title interests. Where the rights provided by the title were not in conflict, it was possible for native title rights to be exercised.

Pastoralists perceived threats to the security of their tenure and reacted strongly to the decision, arguing that coexisting native title compromised lease-holders' capacity to use leases in the manner intended. The Howard Government responded by introducing amendments to the *Native Title Act* to unambiguously protect the rights of leaseholders to undertake existing activities. These amendments took effect in 1998.

The revisions provide for diversification of pastoral activity to include agriculture, forestry, aquaculture and tourism—without attracting the 'right to negotiate'. Such diversification and associated developments require acquisition of native title rights and compensation, the costs being met by Government (the public) rather than the leaseholder. Some constraint was placed on the extent of diversification on larger leases (>50 sq. km), so that no more than 50% of the lease could be used for non-pastoral purposes. For clarity, a note was added to a relevant section of the legislation stating that native title holders are subject to laws of general application, a position which was the law in any event. For example, if all persons were prohibited from lighting fires, the native title holders would also be subject to this general prohibition. How else might these provisions relate to the use of fire?

Aboriginal people in the Australian savannas describe the use of fire as integral to their lives and normal activity. Constraints on rights to use fire therefore would constrain many other activities that are integral to any meaningful exercise of native

A private pastoral perspective

Clearly Native Title legislation has added to the complexity of managing leasehold land in Australia. Until the laws are clarified over the next decade or two, implications for individual leases may be quite different, and discussions of its implications will not be straightforward..

However the following general facts are pertinent to the issue.

- Fire is critical to the sustainable management of the majority of pastoral leases in the tropical savannas.

- Equally the prevention of unplanned or unwanted wildfires is a critical priority aspect of station management to which pastoral enterprises devote considerable financial and management resources.

Managing the Risk of Fire

Fire represents one of the risks, if not the greatest risk of short term financial ruin of pastoral enterprises. For this reason alone I cannot imagine any pastoral manager or the industry at large delegating or negotiating their ability to properly manage this risk. Protection of pasture resources, livestock welfare and economic viability are the paramount concerns that will be little affected by title rights, including maintenance of spiritual connections with land. The extent to which such considerations might ultimately affect use of fire in the exercise of native title rights will depend on the detail of legislation covering fire management and the circumstances in which use of fire is sought. For example, law relating to fire often provides options for its use under a range of circumstances and to meet a range of land management objectives (e.g. to assist in mustering operations), rather than a general prohibition.

Most land managers in northern Australia employ fire as an important tool and seek considerable discretion in its use. Rather than being caught in some general prohibition, customary use of fire on areas where Aboriginal people have native title rights might be influenced by arguments about the compatibility of that use with the activities for which statutory title has been issued. Further, when deciding whether to make an order which regulates or prohibits fires, a Minister (or other relevant person) might in some cases—if an Aboriginal group is especially dependent for sustenance on using fire, for example—have to consider what effect such an order may have on Aboriginal foraging practices. Depending on the circumstances, a failure to at least consider the issue might be an error of law and capable of being reviewed.

Elsewhere in this book we have summarised the importance of fire use for maintaining biodiversity as an essential component of sustainable land management. One of the most important observations is that there is no single prescription to produce the best results for wild plants and animals. Diverse fire regimes are necessary to produce the varied landscapes that support diverse wildlife populations. Patterns of fire use developed by Aboriginal people can play an important role in management for biodiversity.

The existence of native title rights clearly adds another layer of complexity to the already complicated choices that land managers must make about fire use and management. But accommodating multiple views of acceptable practice and seeking management solutions that integrate different perspectives is an obligation that comes with the globalisation that characterises contemporary economic and production systems.

The *Native Title Act* provides mechanisms for reaching and formalising agreements about relationships between native title rights and other rights in an area. Such agreements are likely to become an important part of the social and resource management landscape of northern Australia, and this may include the way in which we manage fire.

by Peter Whitehead

accommodating multiple viewpoints on acceptable fire practice. Good fire defenses require good planning and preparation and the ability to respond to crisis situations with absolute surety and management control.

Fire in Management

Good practice in fire management is surely based on a good understanding of the system and the implications of various management strategies. This knowledge can obviously come from a number of sources and approaches, but I can think of very few successful examples where legislation has actually been anywhere near as effective (if at all) as education in improving the current situation. It

is highly unlikely that the Native Title legislation (introduced for entirely different purposes) will be a catalyst for improved fire management or sustainablity of enterprises and properties in the tropical savannas, whether they be pastoral leases, or any other affected forms of land tenure. However in any potential conflict between rights of pastoral lessees and native title holders it is hard to think of a clearer case in which the right of the pastoral lessee must prevail, than that in which the ability of the pastoralist to protect and manage the resource and assets which underpin their enterprise is threatened.

by Tom Stockwell

The role of technology

Technology cannot help set the goals for land management, but it can help in developing land management practices to achieve those goals and monitor how well we are doing. For example, it will be no easy task to establish what is an appropriate fire regime to take account of all the environmental, ecological, legal, sociological and economic issues confronting even a small part of north Australia. This increasing complexity is making the choice of optimal fire management strategies more difficult. However, the tools to consider such a range of issues are also improving. The challenge is to collate existing and new information in ways that can best assist land managers across the diverse environments and ecosystems of the north. Computer and information technology will be crucial to meeting that challenge. Recent advances in understanding and predicting fire behaviour, fire impacts and fire occurrence all rely on modern computer technology. However, work is still need to develop these into practical tools to assist land managers make sound decisions about fire management.

Fire behaviour

Research into patterns of fire spread over the past two decades has enabled much better predictions of fire behaviour. The spread of fire under different conditions can be simulated with considerable accuracy using computers. However, the practical application of such fire spread models is currently restricted by the limited availability or expense of digital terrain maps and insufficient data on the typical wind behaviour expected throughout the year across the vast landscapes of north Australia. Further development of fire behaviour models could improve the on-ground management of fires.

Smoke management

One of the major public concerns about the use of fire is smoke management. Predicting the spread of smoke from fires using atmospheric circulation models has helped maintain public support by allowing fire strategies that reduce the movement of smoke towards cities to be developed. Improving this predictive capability and using it to plan burning operations is a critical issue in many areas where human populations can be affected by smoke.

Fire impacts

The impacts of fire on native fauna and flora are well documented for some sites in north Australia. Our ability to extrapolate that knowledge to other sites and to the range of possible fire regimes is still in its infancy. Computer models are now providing a way of carrying out this task. For example, the recently developed FLAMES model integrates existing knowledge on the impacts of fire and climate on populations of savanna trees. Adding information

D Dildine DPI

on the dynamics of grass and animal populations could enable this model to investigate the long- and short-term effects on savanna ecosystems of different fire regimes. This approach would be particularly valuable where changes in tree density over time is an issue, as with carbon storage. The emissions of greenhouse gases and atmospheric pollutants under different fire manage-ment regimes can also be estimated from such computer models.

Fire occurrence

Satellite remote sensing is the only practical approach to routine monitoring of fires and mapping of fire history at scales appropriate for north Australia. The potential accuracy of fire scar mapping increases with increasing resolution in time and in space of the imagery, but so does the cost. However, the technology is improving and experience in its use is growing.

As the technology to map fire scars improves, land managers will increasingly be able to use this information to see the spread of their controlled fires. Already maps of fire scars across northern Australia can be downloaded from the Internet (see Chapter 7).

Conclusion

In many parts of north Australia, fire is being rediscovered as a land management tool; it will remain the cheapest way of managing pastures and trees in north Australian savannas for the foreseeable future. Nevertheless, its use has long been questioned by people legitimately concerned about its potential impacts on feed availability, as well as on life and property. Increasingly, issues of air quality, biodiversity and greenhouse gases are being added to that list.

Savanna burning has an important role across north Australia but, in the face of the increasingly complex array of issues confronting land managers, its use must become increasingly sophisticated. Good communication will be essential. Fire management must be seen by land managers and by the public as a safe, effective and efficient land management tool that does address society's concerns and goals. Fortunately, although the questions for land managers are getting harder, the technology to answer them is improving. The challenge is for fire researchers and land managers to develop those tools so that decisions about the use of fire can take account of all the relevant concerns.

Glossary

aerial control burning—ACB; strategic aerial burning operations undertaken from a helicopter or aeroplane to develop typically linear *firebreaks*, especially useful in remote and rugged terrain

annual plants—plants whose reproductive life cycles are completed within one year

backing fire—fire burning into the wind

back-burn—a *backing fire* often applied in fire control operations as a means for directing a relatively low-intensity fire into the path of an oncoming wildfire

char height—the height of flames indicated by burnt, blackened (i.e. 'charred') leaves that remain on the tree or shrub

curing state—the proportion (%) of brown dead matter to green living matter in the ground cover (grass and litter) *fuel load*

fire behaviour—physical attributes of individual fires: height and depth of flames, *rate of spread*, intensity, size and shape of various burning fronts, and intensity

firebreak—any break, whether constructed or natural, which restricts a fire from spreading further.

fire intensity—the rate at which heat released from a linear section of the fire front usually expressed in units of kilowatts per metre of fire edge, kW/m

fire regime—attributes of fires in any one region over a number of years including: extent, seasonality, frequency, intervals between fires, intensity, patchiness

fire scar—an area of burnt vegetation as mapped from aerial photography or satellite imagery

fire-sensitive species—plant species which are readily killed by fires. These typically comprise *obligate seeder* species, but also include a variety of less fire-hardy *resprouters* which may be susceptible to relatively low-intensity fires

fire suppression—activities involved in the containment and eventual extinguishing of an unwanted fire

fire weather—the combination of climatic conditions important for influencing *fire behaviour*, i.e. temperature, wind speed, relative humidity

flaming combustion—the initial phase of the fire characterised by the rapid combustion of light fuels such as leaves, litter, grasses

flanking fire—typically relatively low-intensity fires burning at right angles to the prevailing wind direction

fuel load—the amount of standing grass and *litter* fuel, usually expressed as oven-dry weight of fuel per unit area (e.g. tonnes per ha; kg per sq. m)

Geographic Information Systems (GIS)—computer-based mapping software used for undertaking often complex landscape-scale assessments

greenhouse gases—includes those gases emitted from fires, such as methane and nitrous oxide, which entrap incoming solar energy and thus enhance the process of atmospheric warming

heading fire—fires burning with the wind

heat yield—the heat output from burning fuels, measured in kilojoules per kilogram of dry fuel

hotspots—sites detected by thermal satellite sensors which are relatively hotter than surrounding areas, such as fire fronts

Landsat imagery—fine resolution satellite imagery (*pixels* 30 m x 30 m) used for fine-scale mapping of *fire scars*

line ignition—ignitions applied (typically with a drip torch) with a continuous line of fire along a fire front

litter—that component of the *fuel load* comprising dead leaves, small twigs, etc.

mesic savanna—savanna vegetation occurring in areas where long-term, mean annual rainfall is greater than 900 mm

mosaic burning—burning with the intention of creating small patches, resulting in a landscape characterised by habitat patches of different fire ages

NDVI (Normalised Difference Vegetation Index)— a measure of the 'greenness' of vegetation, derived from *NOAA-AVHRR* imagery.

NOAA-AVHRR imagery—coarse resolution imagery (*pixels* >1.1 sq. km) used for detecting *hotspots*, and also for mapping of fire scars at the broad regional scale

obligate seeder—plants which regenerate solely from seed held on the plant or in the soil after adults have been killed by fire

perennial plants—plants which live for two or more years

perimeter ignition—line ignitions which are lit on several sides of a burn area so that the fire is drawn into the centre as it develops

pixel—the minimum area detectable by a satellite sensor, e.g. 30 m x 30 m for *Landsat* ETM imagery; 1.1 sq. km for *NOAA-AVHRR* imagery

point ignition—fire ignited from a single point, as in
aerial control burning

prescribed fire—any fire which is lit for management
purposes

progressive burning—the practice of burning areas
progressively throughout the year as fuels
dry out

rate of spread—the speed with which the fire travels,
typically measured as metres per second or
kilometers per hour

resprouter—plants which possess the capacity to
resprout from dormant buds on stems or
from root bases following a fire

riparian vegetation—plant communities found
within or adjacent to rivers and streams

rotational burning—process of burning different
parts of a paddock or specified area
sequentially over a number of years,
typically using fires of different intensities

scorch height—height above ground which leaves
in the canopy are killed by heat, and thus
'browned'

seed bank—reserve of seeds held either on the plant
or in the soil. Soil seed banks of some
species, especially legumes, may remain
viable for decades

semi-arid savanna—savanna vegetation occurring in
areas where long-term, mean annual rainfall
is less than 900, mm

smoldering combustion—the phase of the fire
following *flaming combustion* where heavier
fuels such as sticks and logs are slowly
consumed

strategic burning—the use of *prescribed fire*, including
aerial control burning, to strategically break-
up paddocks, properties or large regions

tactical firebreaks—includes the use of *back-burning*
and grading operations undertaken in the
course of *fire suppression*

wet season burning—use of *prescribed fire* during the
wet season period to reduce *fuel loads* of
annual grasses, especially *Sorghum* spp.

Index